S0-BLS-401

ADVANCES IN ELECTRON TRANSFER CHEMISTRY

Volume 4 • 1994

ADVANCES IN ELECTRON TRANSFER CHEMISTRY

Editor: PATRICK S. MARIANO
Department of Chemistry and Biochemistry
University of Maryland-College Park

VOLUME 4 • 1994

Greenwich, Connecticut *London, England*

55 Old Post Road, No. 2
Greenwich, Connecticut 06836

JAI PRESS LTD.
The Courtyard
28 High Street
Hampton Hill, Middlesex TW12 1PD
England

ISBN: 1-55938-506-5

Manufactured in the United States of America

CONTENTS

LIST OF CONTRIBUTORS

Beauford W. Atwater III	Bell Communications Research Red Bank, New Jersey
Richard S. Givens	Department of Chemistry University of Kansas Lawrence, Kansas
Patrick S. Mariano	Department of Chemistry and Biochemistry University of Maryland College Park, Maryland
Kevin S. Peters	Department of Chemistry and Biochemistry University of Colorado Boulder, Colorado
Franklin D. Saeva	Corporate Research Laboratories Eastman Kodak Company Rochester, New York
Jean-Michel Savéant	Laboratoire d'Electrochimie Moléculaire de l'Université de Paris Unité Associeé au CNRS Paris, France
Ung Chan Yoon	Department of Chemistry College of Natural Science Pusan National University Pusan, Korea

PREFACE

The consideration of reaction mechanisms involving the movement of single electrons is now becoming quite common in the fields of chemistry and biochemistry. Studies conducted in recent years have uncovered a large number of chemical and enzymatic processes that proceed via single electron transfer pathways. Still numerous investigations are underway probing the operation of electron transfer reactions in organic, organometallic, biochemical, and excited state systems. In addition, theoretical and experimental studies are being conducted to gain information about the factors that govern the rates of single electron transfer. It is clear that electron transfer chemistry is now one of the most active areas of chemical study.

The series, *Advances in Electron Transfer Chemistry*, has been designed to allow scientists who are developing new knowledge in this rapidly expanding area to describe their most recent research findings. Each contribution is in a minireview format focusing on the individual author's own work as well as the studies of others that address related problems. Hopefully, *Advances in Electron Transfer Chemistry* will serve as a useful series for those interested in learning about current breakthroughs in this rapidly expanding area of chemical research.

Patrick S. Mariano
Series Editor

INTRAMOLECULAR PHOTOCHEMICAL ELECTRON TRANSFER (PET)–INDUCED BOND CLEAVAGE REACTIONS IN SOME SULFONIUM SALT DERIVATIVES

Franklin D. Saeva

Advances in Electron Transfer Chemistry
Volume 4, pages 1–25.

ISBN: 1-55938-506-5

1. INTRODUCTION

The photochemical behavior of sulfonium and onium salts, in general, has received increased attention in the past decade primarily due to applications that use the photochemically generated Brönsted acid in photopolymerization[1] and photoresist technologies.[2] Applications that use photoacids in phototherapy, drug delivery, and protonic conductors are at a much earlier stage of development. Scientific interest in onium-salt photochemistry, on the other hand, originates from the fact that direct photochemical excitation together with electron transfer sensitization produces bond cleavage and high-energy radical, carbocation and ion–radical species.[3] These species undergo secondary chemistry to generate a Brönsted acid that is useful in the technological applications previously mentioned.

This chapter describes our investigations of the photochemical behavior of two classes of sulfonium salt derivatives. In the first class the sulfonium moiety is directly attached to the light-absorbing chromophore and modifies its electronic absorption properties. In the second class, on the other hand, the sulfonium group is separated from the light-absorbing chromophore by a linking group. In this class the sulfonium group does not significantly modify the basic light-absorbing properties of the chromophore.

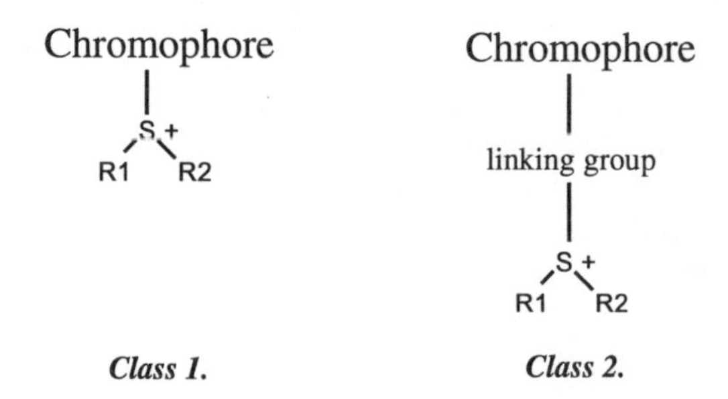

Class 1. *Class 2.*

The basic concepts of photoinduced electron transfer will be described in terms of molecular orbitals (MOs) prior to a discussion of the photochemical behavior of class 1 and 2 sulfonium salts. Emphasis will be placed on the mechanistic aspects of the photochemical reactions as well as on ways as to control PET bond cleavage efficiency and excitation wavelength.

Photochemical excitation of a chromophore involves the promotion of an electron from a filled MO to an empty MO resulting initially in a short-lived singlet excited state. This diradical species has two MOs that are singly occupied (SOMO). More specifically, the longest wavelength electronic transition in a molecular species involves the movement, to a first level of approximation,[3] of an electron from the highest occupied MO (HOMO) to the lowest unoccupied molecular orbital (LUMO) of the light-absorbing chromophore. We will describe molecular systems in which the photoexcited electron initially occupies a π^* MO that lies at higher energy than the σ^* LUMO of the molecule.

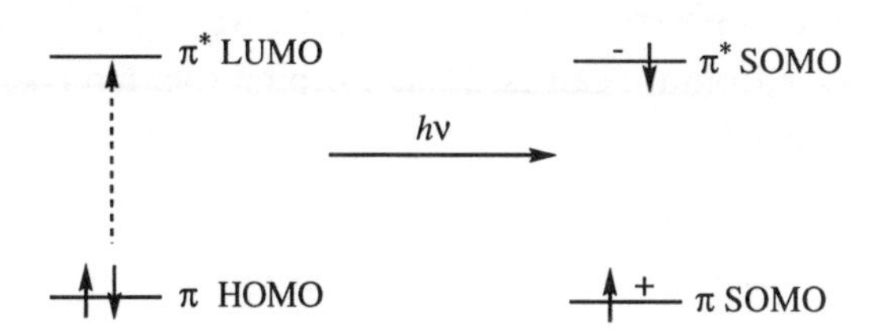

As a direct consequence of photochemical excitation, the molecular species is both a better oxidizing and a better reducing agent. The lowest energy MO available for electron acceptance (i. e., for oxidation of a secondary species), is lower in energy by approximately the photon energy $h\nu$. In addition the photoexcited electron for a molecule in the S_1 excited state is more reducing by approximately $h\nu$. One would clearly expect to observe more excited-state than ground-state electron transfer (ET) reactions.

The free energy change ΔG^o_{ET} for photoinduced electron transfer from an electron donor D or to an electron acceptor A can be determined from the Weller equation:[4]

$$\Delta G^o_{ET} = E^{o'}_{ox} - E^{o'}_{red} - \Delta^1 E_{0,0} + C$$

where

$E^{o'}_{ox}$ = oxidation potential of D
$E^{o'}_{red}$ = reduction potential of A
$\Delta^1 E_{0,0}$ = singlet energy or triplet energy ($\Delta^3 E_{0,0}$)
C is a Coulomb electrostatic term.

In this chapter we will be limiting our discussion to systems that exhibit intramolecular PET.

In a typical organic molecule, the photoexcited electron generally resides in either a π^* or σ^* MO. When the LUMO is σ^* in nature, bond cleavage is possible, provided the rate of back electron transfer is slow relative to the rate of bond cleavage.

In most of the organic molecular systems to be discussed, the LUMO is σ* in nature and transfer of an electron to that MO leads to bond cleavage, provided again that the rate of bond breaking is faster than back electron transfer to regenerate the molecular ground state. In class 1 salts the LUMO of the light-absorbing chromophore is σ*, and orbital overlap between the π* MOs and σ* MOs can be readily achieved by rotation around an aryl–S single bond. In class 2 sulfonium salts the LUMO of the isolated chromophore is π*, while the LUMO of the chromophore attached to the sulfonium salt is σ* and orbital overlap (i.e., σ*/π*) is not readily achieved.

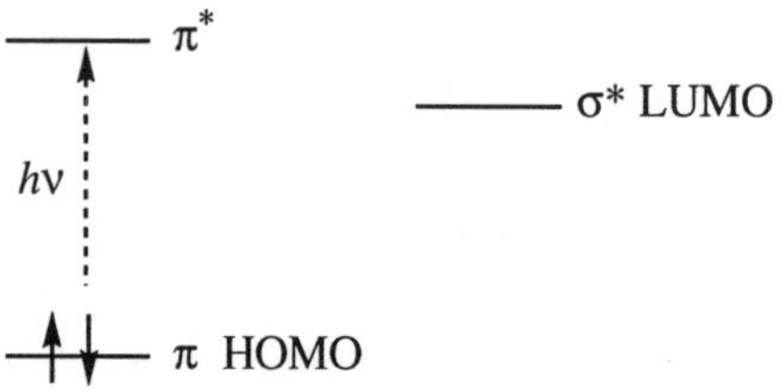

π*

σ* LUMO

$h\nu$

π HOMO

Class 1 structures. Orbital overlap between π* and σ*.
Class 2 structures. No clear overlap between π* and σ*.

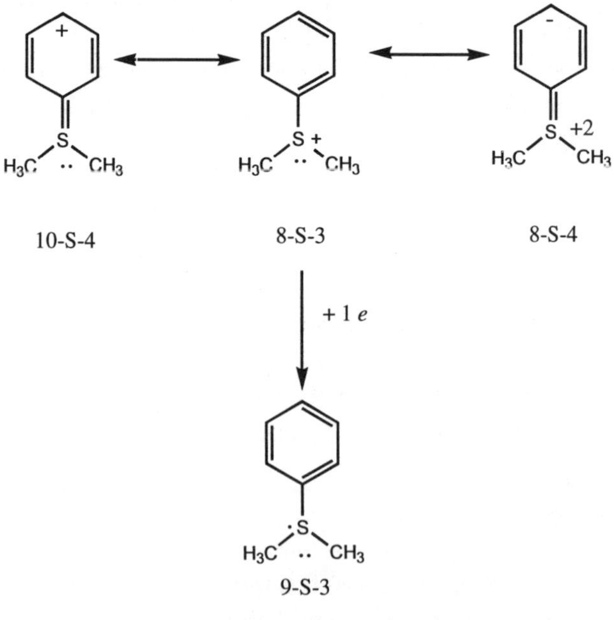

H_3C CH_3 S

10-S-4

H_3C CH_3 S +

8-S-3

H_3C CH_3 S +2

8-S-4

+ 1 *e*

H_3C CH_3 S

9-S-3

Sulfuranyl radical

The sulfonium group is an extremely electron-withdrawing functionality through both inductive and resonance effects. Its Hammett σ_p and σ_m values are more positive than those of substituents such as nitro, methylsulfone, cyano, and trimethylammonium.[5] An aryl ring can donate an electron pair by resonance to the sulfonium group to provide a 10-S-4 system. In addition, the electron pair on S can be delocalized into an adjacent phenyl, for example, to provide an 8-S-4 system, by the classification scheme proposed by Perkins et al.[6]

There is considerable interest in the existence of sulfuranyl radicals formed by one-electron reduction of sulfonium salts.[7] The stability of sulfuranyl radicals as a function of chemical structure is an important scientific issue, and some evidence related to the existence of these species will be presented.

The electrochemical behavior of sulfonium salts within a series of phenyl- and naphthylmethyl-substituted alkylsulfonium salt derivatives was investigated as a function of the substituent group.[8] The electron-withdrawing behavior of the substituent was found to alter the irreversible E_p value by ~1.5 eV, indicating that the reductive cleavage for these species may be concomitant with electron acceptance.[8]

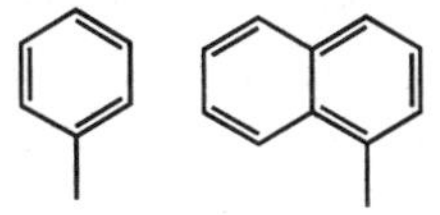

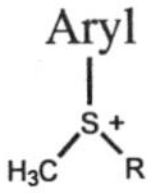

Aryl–S^+(H_3C)–R

R	E_p^c, eV (phenyl)	E_p^c, eV (1-naphthyl)
CH_3	–1.64	–1.51
$CH(CH_3)_2$	–1.61	–1.49
$H_2CC_6H_5$	–1.20	–1.23
$H_2CC_6H_4$-*p*-CN	–0.95	–0.92
$H_2CCOC_6H_5$	–0.73	–0.74
$H_2CC(C_6H_5)=C(CN)_2$	–0.23	–0.17

Aryl = phenyl, 1-naphthyl

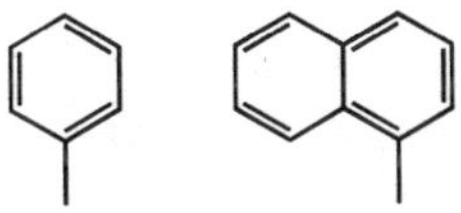

The experimental evidence suggests that the one-electron reduced sulfonium salt does not form a stable sulfuranyl radical intermediate that eventually undergoes bond cleavage. The formation of a common one-electron reduced intermediate throughout the series would result in a change in E_p^c of ~0.3 eV.[8] It would be reasonable to assume that sulfuranyl radicals can be viable intermediates provided that the proper substituent groups are present. The electrochemical evidence suggests that as the electron is added to the lowest energy σ*, molecular orbital cleavage of that C–S bond cleavage occurs within the time scale of a molecular vibration to provide a carbon-based radical, as shown below.[8]

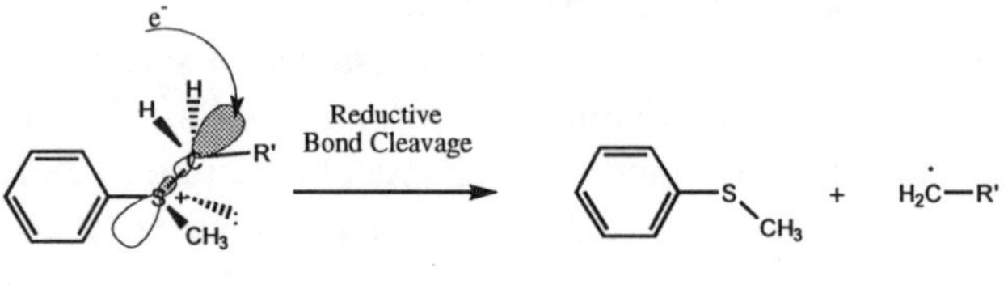

Reductive Bond Cleavage

σ* LUMO

The photochemical behavior of class 1 and class 2 sulfonium salts will be presented in the following section, starting with ultraviolet light–absorbing chromophores. The effect of transferring an electron from the HOMO to the σ* LUMO by photochemical excitation will be discussed.

2. DISCUSSION

2.1. Class 1 Sulfonium Salts

Substitution of a dialkyl sulfonium moiety on the 1-position of naphthalene shifts naphthalene absorption to the red by ~20 nm in acetonitrile.[9] Irradiation of a 1-naphthylmethylsulfonium salt derivative in acetonitrile solvent at wavelengths greater than 310 nm produced four products including a Brönsted acid when R′ = *p*-cyanophenyl.[9]

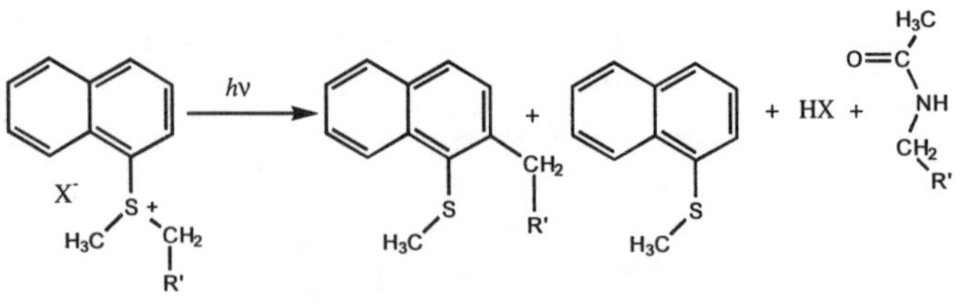

1 **2** **3** **4**

The photofragmented substituted alkyl group in the sulfonium salt couples with 1-thiomethylnaphthalene to form **2** and with solvent to form the corresponding amide **4**. The photoreaction has been shown to involve the photoexcited singlet excited state of 1. We have proposed that a singlet cation–radical/radical pair intermediate is involved in the photochemical transformation:

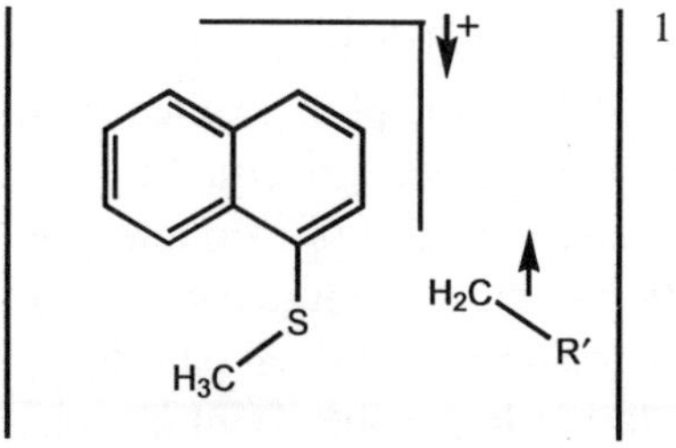

Coupling of the radical with the cation–radical at the 2-position results in **2** after proton loss. Compound **3** is formed, we believe, as a result of electron transfer from the radical to the naphthalene cation–radical in systems where electron transfer is thermodynamically favorable. The carbocation formed reacts with acetonitrile reversibly to form amide with trace water. Another possible pathway for the formation of **3** and **4** is nucleophilic attack by solvent on the *p*-cyanobenzyl group, for example, in the dihydro-adduct. When X = BF_4^-, tetrafluoroboric acid is formed with a quantum yield of 0.34 for R′ = *p*-cyanophenyl. The quantum yield for the formation of **2** and **3** is 0.18 and 0.16, respectively, in this case.

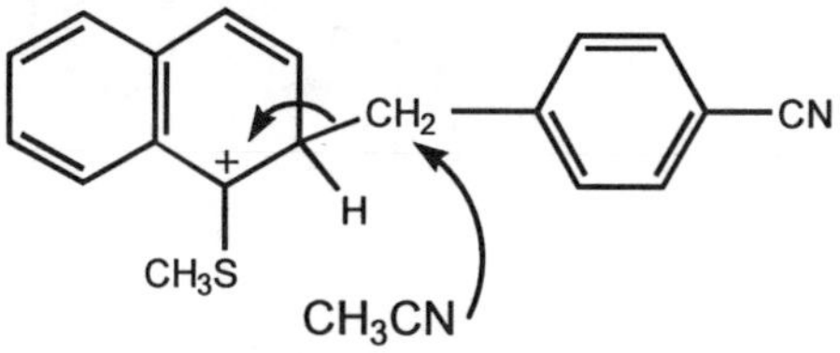

In class 1 naphthalene sulfonium salts the LUMO is clearly σ* from AM1[10] molecular orbital calculations. The σ* LUMO can overlap with the π-orbitals on naphthalene for efficient transfer of the photoexcited electron as shown.

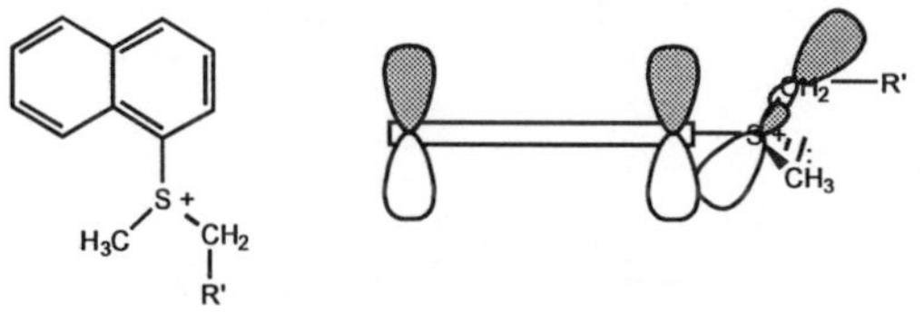

Support for the presence of the intermediate singlet cation–radical/radical pair was obtained from an investigation of the photochemistry of some Group V onium salts.[11] An ammonium, phosphonium, and arsonium salt was irradiated in argon-purged acetonitrile solvent and the photoproducts investigated. Based on the observed photoproducts it was concluded that the intermediate singlet cation–radical/radical was formed from the photoexcited singlet state in each case and the product distribution was governed by the nuclear spin of the heteroatom in the cation radical. For example, the nitrogen- and phosphorus-centered cation–radical undergo more rapid intersystem crossing to form the triplet cation–radical/radical pair than in the arsenic derivative because of the larger hyperfine coupling constant. As a result the ammonium and phosphonium salts provide only out-of-cage products such as a bibenzyl derivative and a product derived from H-atom abstraction in the presence of an H-atom source such as thiophenol.[11] The arsonium salt, on the other hand, provided a mixture of out-of-cage and in-cage photoproducts due to the smaller hyperfine coupling constant for As. In comparison, the sulfonium salts provide only in-cage coupling, since S does not have a nuclear spin.

out-of-cage products
ISC
major pathway when X = P
in-cage coupling
major pathway when X = As

When the aryl group in aryl sulfonium salts is 9-anthryl, the electronic absorption spectrum is red shifted by ~40 nm from that of 2-thiomethylphenylanthracene in acetonitrile.[12] The short-axis in-plane ^1L_b electronic transition is red-shifted and broadened compared to the long-axis in-plane ^1L_a electronic transition. There is a long-wavelength shoulder at ~420 nm in acetonitrile that we have assigned to the π to σ^* charge transfer transition.[11]

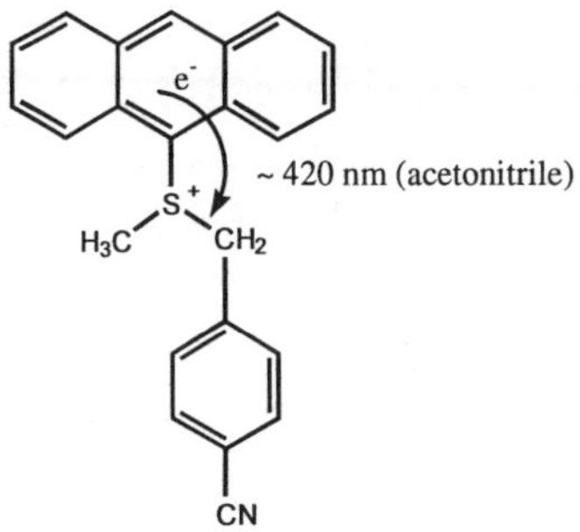

e⁻
~ 420 nm (acetonitrile)
H_3C S⁺ CH_2
CN

5

Irradiation of **5** with light of a wavelength greater than 400 nm in argon-purged acetonitrile resulted in rapid bleaching of the absorption in that spectral region. Five regioisomers (**6**) were obtained in 56% yield (0.43 quantum yield) in addition to photosolvolysis products (**7, 8**) in

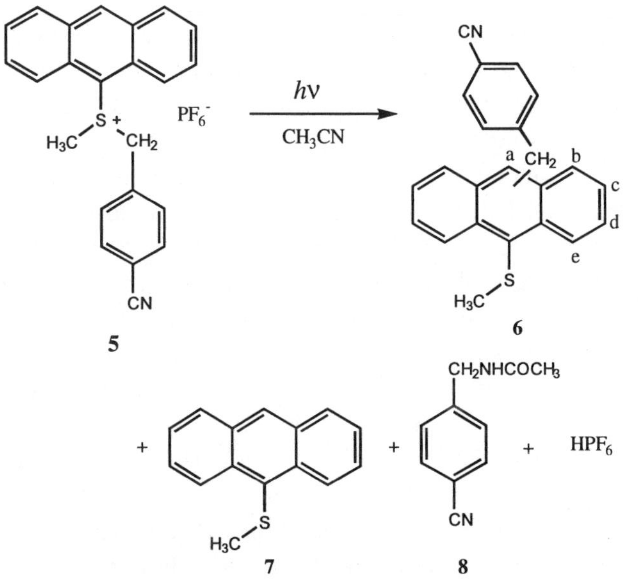

H_3C S⁺ CH_2 PF_6^- CN **5**
$h\nu$ / CH_3CN →
CN, a CH_2 b, c, d, e, S, H_3C **6**

+ S, H_3C **7** + $CH_2NHCOCH_3$, CN **8** + HPF_6

44% yield (0.34 quantum yield). The predominant rearrangement product is the 10-*p*-cyanobenzyl-9-thiomethylanthracene in 38% overall yield.[12]

Regioisomeric photoproducts **6**a–e had the following distribution: 68% **6**a, 5% **6**b, 2% **6**c, 10% **6**d, 15% **6**e. The high percentage of **6**a can be accounted for either by electrophilic substitution of the benzyl carbocation on thiomethylanthracene or by radical/radical–cation coupling to produce an *ipso*-substituted intermediate that rearranges to the final product. The intermediate singlet cation–radical/radical pair can in theory couple at all sites that have spin density.[12] A comparison of the spin densities in the 9-thiomethylanthracene cation–radical, from AM1[10] molecular orbital calculations, with the product distribution in **6**a–e, clearly demonstrates that radical/cation–radical coupling is a kinetic rather than a thermodynamic process.

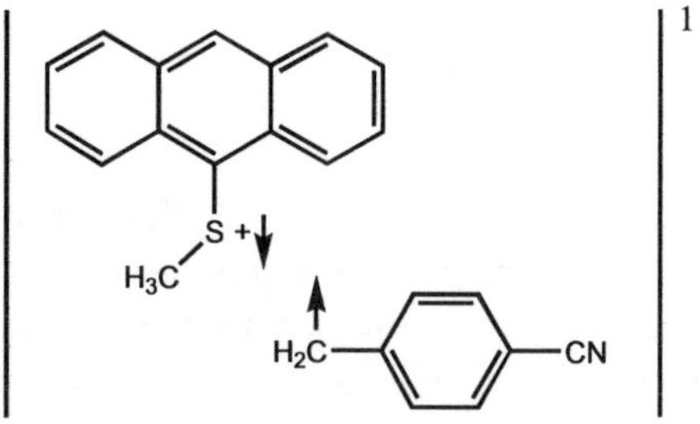

In the anthryl sulfonium salt **5**, the LUMO is again σ* as indicated by AM1 MO calculations. The π* MO of anthracene lies above (i.e., at higher energy than) the σ* LUMO, which is localized predominantly on the sulfonium group.

AM1 MO calculations[10] were performed on a series of arylmethyl-*p*-cyanobenzyl sulfonium salts where aryl was varied from phenyl, 1-naphthyl, 9-anthyl, to 5-naphthacenyl.[13] The results indicate that as the number of condensed aromatic rings increases the energy of the π HOMO level increases, while the energy of the π* level decreases in energy throughout the series, as expected. The σ* LUMO, on the other hand, increases in energy as the aryl ring system increases in size, until for the naphthacene case there is a crossover between the π* and σ* MOs where the π* MO lies below the σ* MO. The conclusions from the calculations are supported by experiment. Singlet lifetimes are short and fluorescent quantum yields low for the phenyl, naphthalene, and anthryl sulfonium salt. In contrast the naphthacenyl sulfonium salt exhibited a higher fluorescent quantum yield by a factor of 20 to 100 and a longer singlet lifetime by a factor of 10. The lack of photochemistry in the naphthacene derivative was consistent with the results from calculations as well. These observations support our view that a σ* LUMO is a necessary requirement for PET bond cleavage.[13]

2.2. Class 2 Sulfonium Salts

It became clear that the way to achieve long wavelength–sensitive photoacids was to combine a sulfonium group with a chromophore through an electronically insulating linking group. In this scheme, the σ* LUMO of the sulfonium group must be at lower energy than the π* LUMO of the chromophore in order for the thermodynamics for electron transfer from the photoexcited chromophore to the sulfonium moiety to be exothermic.[14] In addition, it would be desirable to have the π* MOs be in close proximity to the σ* MOs for efficient through-bond or through-space electron transfer.

Phenylanthracene derivatives were thought to possess the desired requirements and to provide a model for the concept and for longer wavelength systems. The phenylanthracene sulfonium salt derivative **9** was synthesized and found to possess a ground state conformation in the

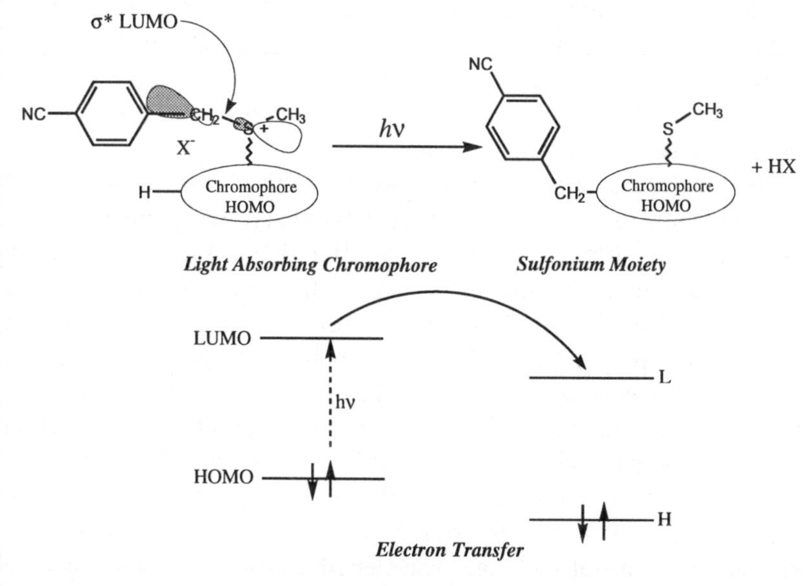

Scheme 1.

solid state, from its x-ray crystal structure, which placed the *p*-cyanobenzylsulfonium moiety over the anthracene ring system in close proximity to its π and π^* MOs.[14] The electronic absorption spectrum of **9** in acetonitrile is virtually identical to that of 9-phenylanthracene except for a modest 5-nm shift to the red.

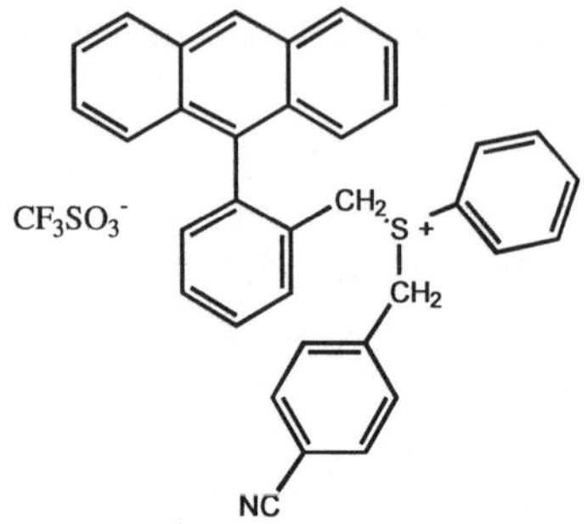

$CF_3SO_3^-$ CH_2 S + CH_2 NC

9

Compound **9** is relatively nonfluorescent with a fluorescence quantum yield of less than 10^{-3}. The lack of fluorescence and short singlet lifetime is consistent with the introduction of a rate process such as electron transfer, which competes effectively with fluorescence and nonradiative decay of the phenylanthracene chromophore. Photoinduced electron transfer from the anthracene ring system to the sulfonium moiety is highly exothermic in acetonitrile, with $\Delta G^{o}_{ET} = -24$ kcal/mol. Photolysis of **9** in acetonitrile (i.e., $h\nu > 360$ nm, so that only the anthracene is electronically excited) provided both rearrangement and products derived from heterolytic bond cleavage.[14]

Rearrangement products **10**a–e were produced with a 44% yield (quantum yield of 0.15).[14] The regioselectivity in rearrangement product **10** (**10**a, 19%; **10**b, 25%; **10**c, 17%; **10**d, 12%; **10**e, 22%) clearly demonstrated either a kinetic coupling of the benzyl radical with the anthryl cation–radical or electrophilic substitution involving the benzyl carbocation. The products of heterolytic bond cleavage (i.e., homolytic bond cleavage and electron transfer) **11** and **8** were produced with a 56% yield and a quantum yield of 0.17. The quantum yield for the production of trifluoromethanesulfonic acid was 0.32. The photoproducts are again consistent with the singlet cation–radical/radical intermediate produced by intramolecular electron transfer. Now that the viability of the scheme has been demonstrated, longer-wavelength absorbing systems can be designed and synthesized.

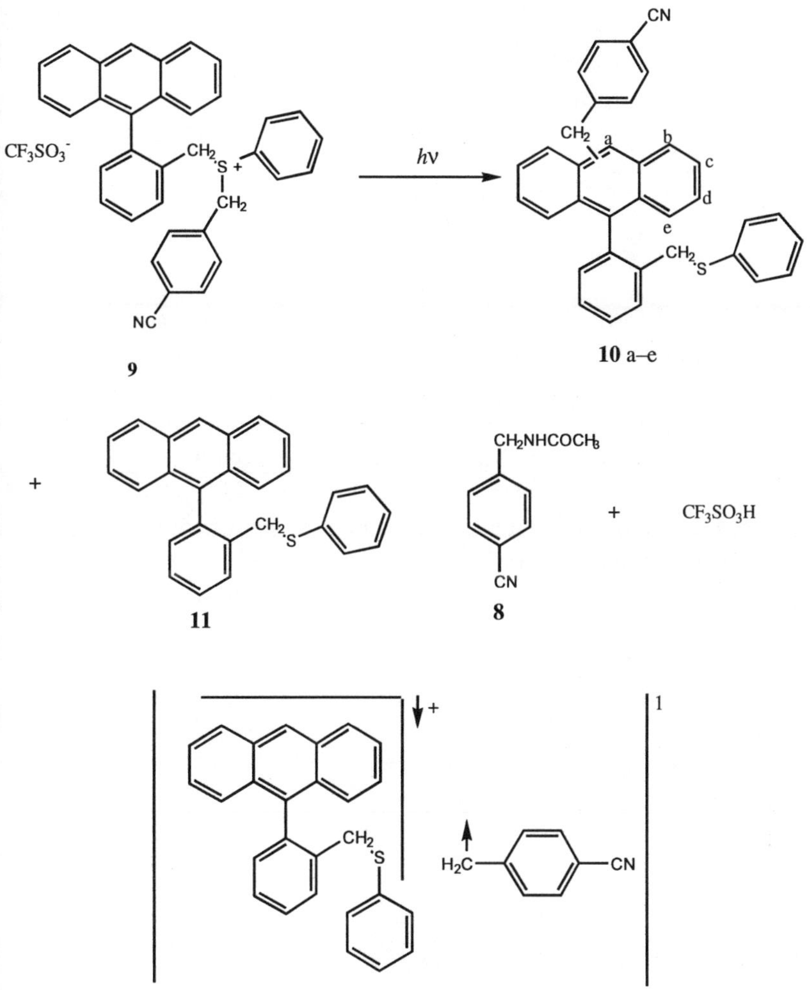

The corresponding (5-phenylnaphthacenyl)sulfonium salt **12** was synthesized. Compound **12** absorbs out to ~520 nm in CH_2Cl_2 and exhibits fluorescence with a quantum yield of 0.25. The absorption spectrum for **12** is similar to the electronic absorption spectrum of 5-[2-[(phenylthio)methyl]phenyl]naphthacene.

The free energy change for PET from naphthacene to the sulfonium moiety is –15 kcal/mol in acetonitrile, approximately 9 kcal/mol less favorable than for **9**. Photolysis of **12** in argon-purged acetonitrile provided photoinduced rearrangement products in quantitative yield with a quantum yield of 0.25.[14]

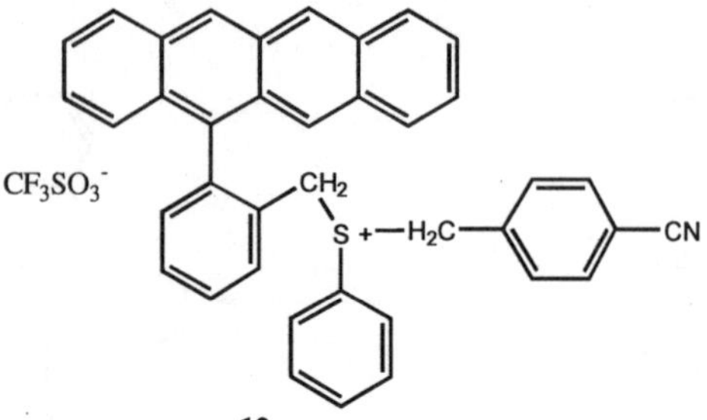

$CF_3SO_3^-$ CH_2 S + — H_2C — CN

12

The lack of formation of the unsubstituted sulfide and amide product from heterolytic bond cleavage, in this case, is attributed to the inability of the phenylnaphthacene cation radical to oxidize the *p*-cyanobenzyl radical. The *p*-cyanobenzyl radical, with an oxidation potential of 1.03 V in acetonitrile (versus SCE), is capable of being oxidized by the cation radical of the phenylanthracenyl sulfide derivative **11** (E_p^{ox} = 1.04 V) and oxidized less readily by the cation radical of the phenylnaphthacenyl sulfide derivative (E_p^{ox} = 0.78 V).

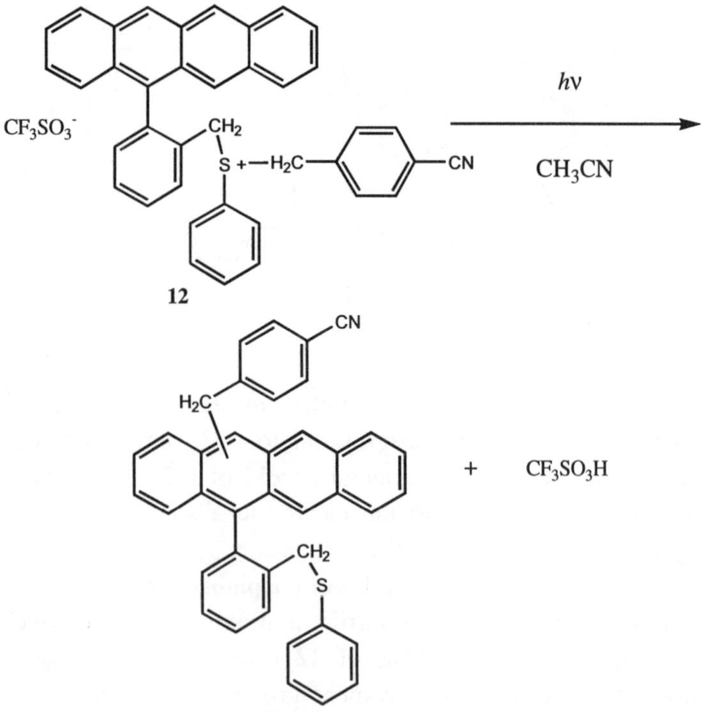

$CF_3SO_3^-$ CH_2 S + — H_2C — CN $\xrightarrow[CH_3CN]{h\nu}$

12

CN H_2C CH_2 S + CF_3SO_3H

13

All eleven possible regioisomers of **13** appear to be formed in the rearrangement process. A comparison of the product distribution with AM1-calculated spin densities in the phenylnaphthacenyl cation radical further demonstrates that the product distribution is determined by spin densities and is kinetically controlled. We do not believe that **13** arises from electrophilic substitution of the benzyl carbocation on naphthacene because amide product is not observed unless electrophilic substitution is considerably faster than reaction with the solvent.

We believe that the efficacy of the photoinduced bond cleavage is due in part to the apparent concerted nature of the bond cleavage reaction. This process normally competes effectively with energy-wasting back electron transfer from the reduced acceptor (sulfuranyl radical) to the oxidized donor (aryl cation radical). Thc close proximity of the aryl π^* and σ^* LUMO sites on the sulfonium moiety would tend to enhance the efficiency for both PET concerted bond cleavage and radical/radical coupling as well.

What happens to the efficiency of PET when the driving force is thermodynamically more favorable? In addition, how does the rate of PET change with the distance between D and A? These questions will be addressed below.

Chemical structures of linked bichromophoric sulfonium salts in the following sections utilize the phenyl linking group. This linking group is still considered insulating since the π conjugation between the naphthacene or anthracene ring and the phenyl is minimized due to the existence of a twisted conformation. When a dimethyl sulfonium moiety is attached directly to the phenyl ring in phenylnaphthacenyl derivatives, the photochemical behavior is very different from that of the methyl-*p*-cyanobenzylsulfonium salt derivatives.[15]

Two series of regioisomeric-substituted 5-phenylnaphthacenylsulfonium salts were synthesized and their physicochemical behavior investigated and compared. The singlet lifetime (τ_{S1}), quantum yield for fluorescence (ϕ_f), quantum yield for acid formation (ϕ_{H^+}), redox behavior, free energy change for PET ($\Delta G^o{}_{ET}$), and electronic absorption behavior were measured for each regioisomer in both series in acetonitrile solvent.

In the dimethylsulfonium trifluoromethanesulfonate salt series (**14**) the wavelength of the lowest energy electronic transition is shifted slightly to the red of that transition in 5-phenylnaphthacene (480 nm, CH_3CN).[15] The singlet lifetime τ_{S1} for the regioisomers varies from 12.9 to 5.5 ns. The singlet lifetime for each of the salts is considerably longer than the 3.7 ns observed for 5-phenylnaphthacene in acetonitrile. The

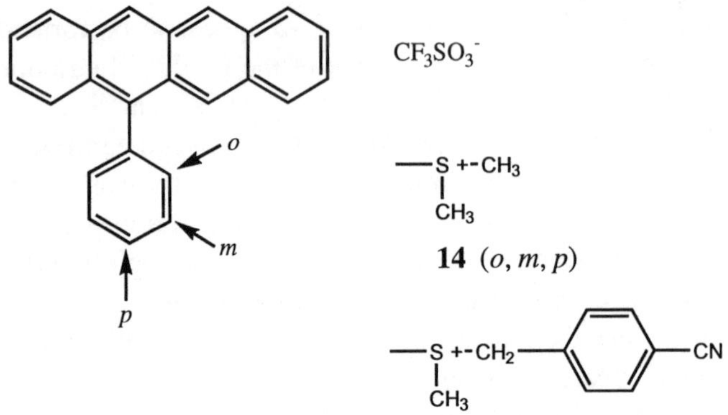

$CF_3SO_3^-$

o

m

p

$-S^+-CH_3$ (CH_3)

14 (*o*, *m*, *p*)

$-S^+-CH_2-C_6H_4-CN$ (CH_3)

15 (*o*, *m*, *p*)

$-S^+-CH_3$ (CH_3) $CF_3SO_3^-$

(CH_3CN)

14

o — τ_{S1} = 12.9 ns (2-DM)
$\phi_f = 0.47$
$\phi_{H^+} < 0.01$
λ_{max} = 486 nm
ΔG^o_{ET} = –1.8 kcal/mol

p — τ_{S1} = 5.9 ns (4-DM)
$\phi_f = 0.34$
$\phi_{H^+} < 0.01$
λ_{max} = 481 nm
ΔG^o_{ET} = –3.5 kcal/mol

m — τ_{S1} = 5.5 ns (3-DM)
$\phi_f = 0.27$
$\phi_{H^+} < 0.01$
λ_{max} = 481 nm
ΔG^o_{ET} = –2.5 kcal/mol

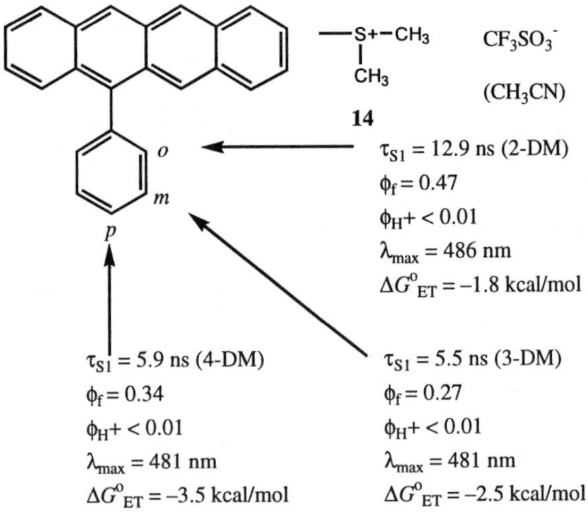

fluorescence quantum yield for **14** varies from 47% to 27%. These values are again considerably larger than that observed for 5-phenylnaphthacene (i.e., $\phi_f = 0.18$) in the same solvent. For each isomer of compound **14**, ϕ_{H^+} is less than 0.01. Since PET is necessary for bond cleavage and Brönsted acid formation, it is clear that the rate of PET is not competitive with the rate of fluorescence and other decay modes for S_1. The rate of fluorescence for 5-phenylnaphthacene is $4.9 \times 10^7\ s^{-1}$ in air-saturated acetonitrile. The thermodynamic driving force for PET is modestly exothermic for each of the regioisomers of **14**, and the average rate of PET is ~$5.6 \times 10^6\ s^{-1}$, considerably slower than the rate of fluorescence.

The variation in τ_{S1} and ϕ_f between **14** and 5-phenylnaphthacene is attributed to the effect of the electron-withdrawing dimethylsulfonium group on the relative energies of the S_1 and T_2 states. The primary pathway for deactivation of the S_1 state for 5-phenylnaphthacene in benzene solution has been found to be intersystem crossing (ISC), with ϕ_{ISC} = 0.65[15]. The rate of ISC from S_1 to T_1 through T_2 is expected to decrease as follows: **14***o* < **14***p* < **14***m* << 5-phenylnaphthacene. The maximum transmission of the inductive effect of the sulfonium group on the naphthacene ring system occurs in **14***o* due to its close proximity. In **14***o* the phenyl ring is expected to be nearly orthogonal to the plane of the naphthacene ring system. AM1 MO calculations indicate the dihedral angle between the phenyl and the naphthacene ring system to be 88°, 72°, and 64° for **14***o*, **14***m*, and **14***p*, respectively.

The effect of increasing the thermodynamic driving force (i. e., ΔG^o_{ET}) for PET by replacing methyl on sulfur with *p*-cyanobenzyl is quite dramatic.[15] The *p*-cyanobenzyl radical is clearly more stable than the methyl radical, and the electrochemical reductive cleavage process for the methyl-*p*-cyanobenzyl sulfonium moiety occurs at a ~0.5V less negative potential than that of the dimethylsulfonium group.

The τ_{S1} for **15***o* is shortened to 0.8 ns from 12.9 ns for **14***o*. In **15***m*, τ_{S1} is unchanged from **14***m*, while τ_{S1} in **15***p* is actually longer than in the corresponding derivative **14***p*. The fluorescence quantum yields for **15** (*o, m, p*) range between 4.2 and 5.2%, considerably lower than that observed for the **14** (*o, m, p*) series. Correspondingly, the quantum yield

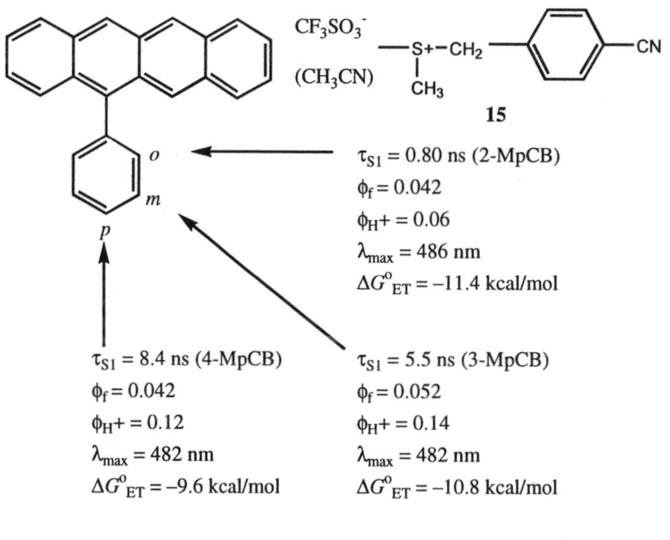

$CF_3SO_3^-$ (CH_3CN) —S^+—CH_2—C_6H_4—CN (CH_3 on S)

15

o: τ_{S1} = 0.80 ns (2-MpCB)
ϕ_f = 0.042
ϕ_{H^+} = 0.06
λ_{max} = 486 nm
ΔG^o_{ET} = −11.4 kcal/mol

p: τ_{S1} = 8.4 ns (4-MpCB)
ϕ_f = 0.042
ϕ_{H^+} = 0.12
λ_{max} = 482 nm
ΔG^o_{ET} = −9.6 kcal/mol

m: τ_{S1} = 5.5 ns (3-MpCB)
ϕ_f = 0.052
ϕ_{H^+} = 0.14
λ_{max} = 482 nm
ΔG^o_{ET} = −10.8 kcal/mol

for acid formation ϕ_{H^+} has increased substantially over that observed for the dimethylsulfonium salt series. The driving force for PET, ΔG^{o}_{ET}, in derivatives of **15**, ranges between –9.6 and –11.4 kcal/mol, depending on the position of substitution.[16]

Comparison of the redox behavior within a series shows the influence of distance between the electron donor and acceptor moieties. The naphthacene ring system exhibits a reversible oxidation potential of ~1.0 V (versus SCE, CH_3CN) and a reversible reduction potential of ~–1.5 V. Reduction of the sulfonium group, on the other hand, is irreversible, with E_p ~–1.5 V for derivatives of **14** and E_p ~–1.0 V for the isomers of **15**.

Photoexcitation of the naphthacene π electrons to a π^* state is the lowest energy electronic transition in both **14** and **15**. For bond cleavage to occur, the photoexcited electron must be transferred to the lower energy σ^* LUMO localized on the sulfonium moiety. The electron transfer process from a π^* to a σ^* state takes place either through bond (i.e., σ or π) or through space. Once the photoexcited electron reaches the σ^* molecular orbital, bond cleavage occurs, provided back electron transfer does not occur at a comparable or faster rate. The π^* molecular orbitals are localized on the naphthacene ring system while the σ^* molecular orbitals are localized on the sulfonium group.

Although the PET process within the **15** (*o, m, p*) series is not precisely isoenergetic in a thermodynamic sense, one can measure the effect of distance on the rate of PET. First, it is assumed that every PET produces bond cleavage and subsequently a proton. The rate of PET (i.e., k_{ET}) is determined from the quantum yield of acid formation and the fluorescent lifetime by the relationship $k_{ET} = \phi_{H^+}/\tau_{S1}$. The assumption that every PET produces bond cleavage is reasonable in view of the fact that the electrochemical reductive cleavage appears to be a concerted electron acceptance and bond cleavage, especially in those cases where stabilized radicals are produced.

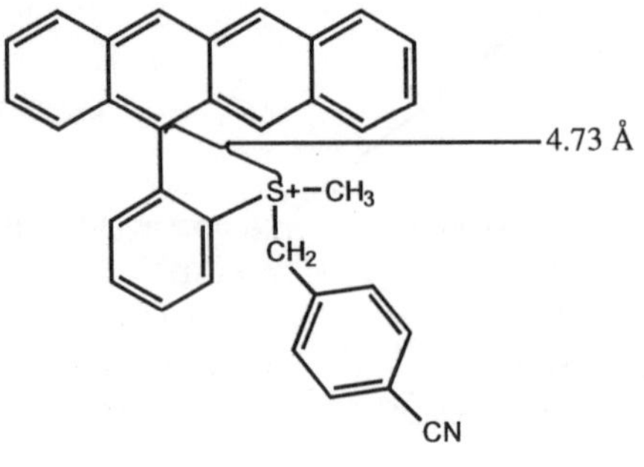

4.73 Å

$S^+–CH_3$

CH_2

CN

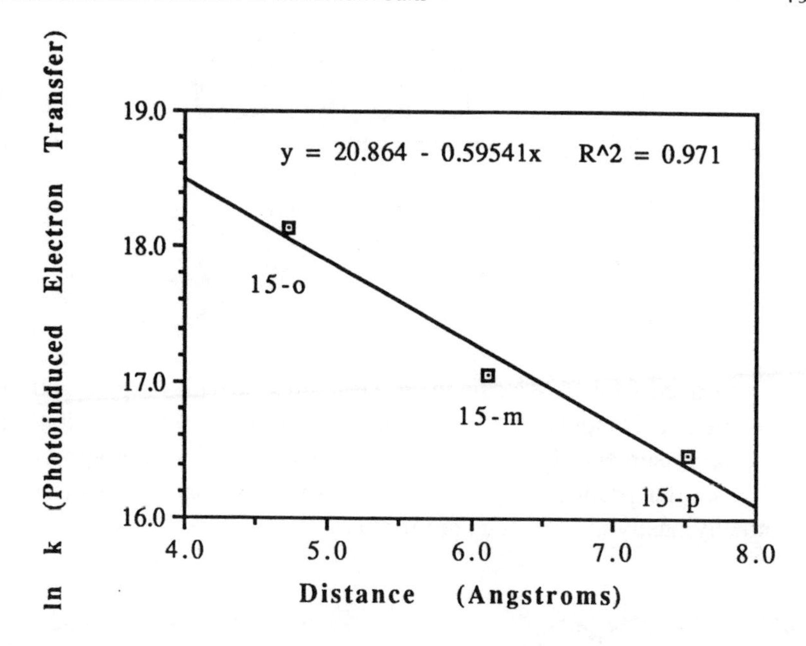

Figure 1. The effect of through-bond distance on electron transfer rate for **15***o*, **15***m*, and **15***p*.

A plot of ln k_{ET} versus through-bond distance in angstroms for **15***o*, **15***m*, and **15***p* is provided in Figure 1. The distance between D and A was taken as the sum of the bond lengths between the naphthalene ring and sulfur.[16] Since the phenyl linking group between the naphthacene ring system and the sulfonium moiety is forced out of plane of the naphthacene ring, it influences electron transfer between the D and A functionalities by mixing its electronic states with those of D and A.

Empirically, the distance dependence of the electron transfer rate constant is given by the relationship

$$\ln k_{ET} = \alpha(r - r_o) + \ln k_o,$$

where k_o is the largest rate constant for electron transfer at contact between D and A, r is the D–A distance, r_o is the distance at van der Waals contact, at which the largest rate occurs, and α is a constant.

The correlation coefficient for the plot shown in Figure 1 is 0.97, with a slope of 0.60.[15] We believe that the correlation shown in Figure 1 is evidence supporting a through-bond rather than a through-space electron transfer mechanism. A plot of ln k_{ET} versus through-space distance taken from the center of the naphthacene ring to the center of the S-CH_2 bond

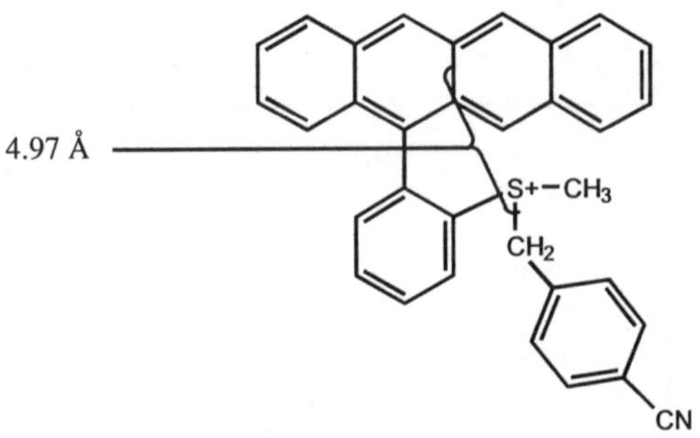

4.97 Å

being cleaved from AM1-calculated structures provides a plot with a correlation coefficient of 0.85 and α of –0.91. Photochemical irradiation of the regioisomers of **15** in acetonitrile produces four photoproducts with a similar product distribution. Both rearrangement products and thiomethylphenylnaphthacene are formed.

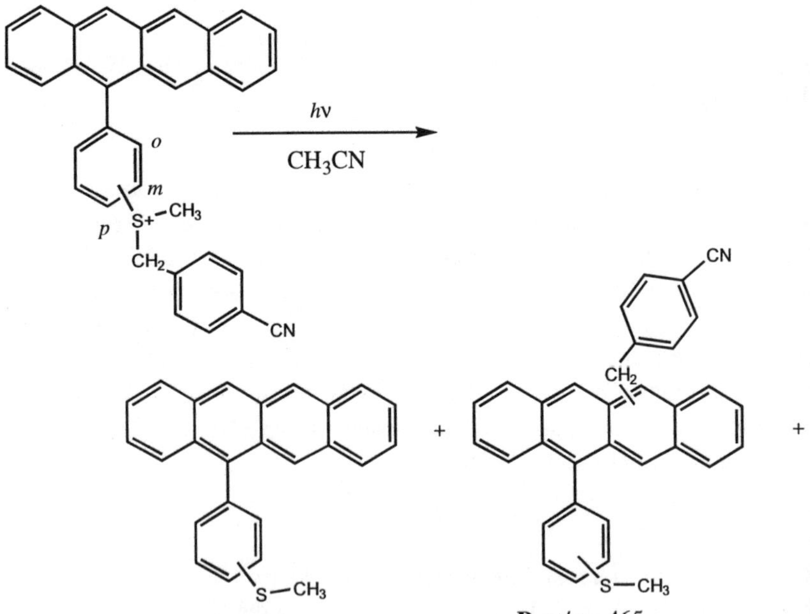

$h\nu$, CH_3CN

A $m/e = 350$

o: 1.0

m: 1.0

p: 1.0

B $m/e = 465$

o: 1.88

m: 2.22

p: 1.91

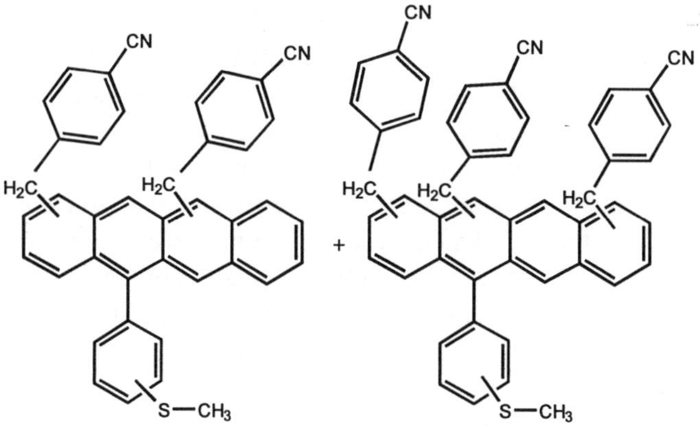

C $m/e = 580$

o: 0.67

m: 1.05

p: 1.27

D $m/e = 695$

o: 0.13

m: 0.38

p: 0.43

Surprisingly, the distribution of photoproducts is essentially distance-independent. The ratio of monosubstituted rearranged product (**B**) to unsubstituted thiomethylphenylnaphthacene (**A**) is ~2 for the three regioisomeric sulfonium salts. In addition to the monosubstituted photoproduct (**B**), disubstituted (**C**) and trisubstituted (**D**) photoproducts are also observed. The existence of the out-of-cage di- and trisubstituted products indicates that the existence of intermolecular PET and/or the *p*-cyanobenzyl carbocation may be involved. The benzylic carbocation intermediate produced by one-electron oxidation of the *p*-cyanobenzyl radical could possibly react with acetonitrile solvent to produce a stabilized electrophilic species. In the absence of nucleophiles such as water, the formation of the acetonitrile adduct is thought to be reversible.

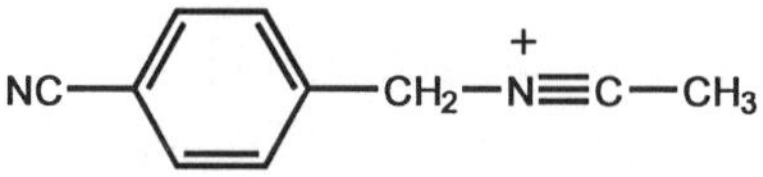

$$NC-C_6H_4-CH_2-\overset{+}{N}\equiv C-CH_3$$

2.3. Class 2 Bis-Sulfonium Salts

Phenylnaphthacene derivatives exhibit considerable ISC from their first excited singlet state to their first excited triplet state. This is an energy-wasting pathway since PET from the singlet excited state leads to bond cleavage and Brönsted acid formation. Both 5, 12-diphenylnaphthacene and 9, 10-diphenylanthracene exhibit a high fluorescence quantum yield and less ISC from the S_1 state than the monophenyl derivatives. With this in mind, bis-sulfonium salt derivatives of 5, 12-diphenylnaphthacene were synthesized and their photochemical behavior investigated. These systems were conveniently synthesized from 5, 12-naphthacenequinone and 9, 10-anthraquinone by a Grignard addition reaction. Compound **16** was synthesized in a three-step reaction sequence.

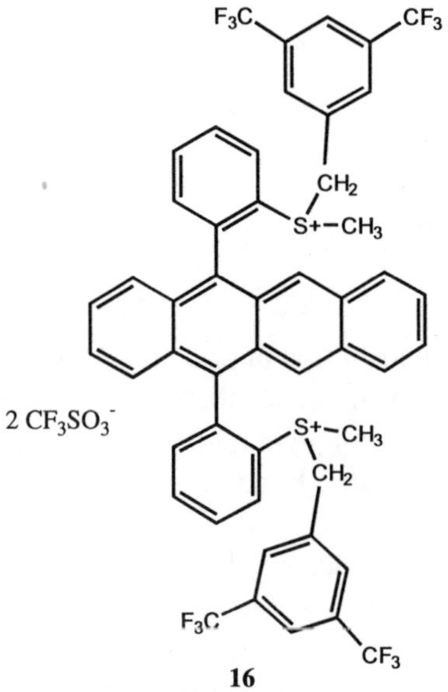

F_3C CF_3 CH_2 S^+-CH_3 S^+-CH_3 CH_2 F_3C CF_3

2 $CF_3SO_3^-$

16

Since the phenyl rings directly attached to naphthacene are twisted out of the plane of the naphthacene ring and the sulfonium groups can lie either above or below the plane of the ring, atropisomerism exists. For example, the 500-MHz ^1H nuclear magnetic resonance spectrum for **16** in deuteroacetonitrile shows four methyl resonances between 3.0 and 3.2 ppm. The benzylic protons are diastereotopic and exist as an AB quartet, and in **16** one observes four sets of AB quartets between 4.7 and 4.9 ppm. The longest-wavelength electronic transition in **16** occurs at 503 nm ($\varepsilon = 7800$) in ethyl acetate. Compound **16** possesses an excited singlet state lifetime of 10 ns and a fluorescence quantum yield of 0.40 in ethyl acetate. Electrochemical analysis of **16** shows two irreversible

$2\ CF_3SO_3^-$

16

$\xrightarrow[CH_3CN]{h\nu}$

$+\ 2\ CF_3SO_3H$

reduction processes: $E_p = -1.12$ V and $E_p = -1.31$ V (50 mVs^{-1} scan rate versus SCE, CH_3CN) for the sulfonium moieties and a reversible reduction at $E^{o\prime}{}_{red} = -1.50$ V for the naphthacene ring system. Electrochemical oxidation of the naphthacene ring is also reversible, with $E^{o\prime}{}_{ox} = 1.23$ V. The free energy change for intramolecular PET from the naphthacene to the sulfonium is exothermic by 2.8 kcal/mol. The quantum yield for formation of trifluoromethanesulfonic acid in acetonitrile was 0.17. The bis-sulfonium salts are uniquely capable of undergoing two sequential PET processes and liberate two molecules of a Brönsted acid for two photons absorbed.

There have been several investigations of the photochemical behavior of ultraviolet light–absorbing diaryliodonium and triphenylsulfonium salts by Crivello,[17–19] Pappas,[20–26] Hacker, [27–33] and Reiser.[34] Attempts to extend the spectral response of these materials were accomplished by intermolecular electron transfer sensitization with electron-rich aromatic compounds such as naphthalene, anthracene, and pyrene. In general, the photochemical behavior of the intermolecular sensitized systems is similar to the intramolecular class 1 and class 2 sulfonium salts, in that radical/ion–radical/carbocation intermediates are formed photochemically.

3. SUMMARY

The photochemical behavior of two classes of bichromophoric sulfonium salt photoacids is presented. Photochemical rearrangement and photosolvolysis products are observed in high yield for both classes through, we believe, a singlet cation–radical/radical pair. Radical/radical coupling provides rearrangement after proton loss, while electron transfer from the radical to the cation radical, within the solvent cage, provides a carbocation that can react with solvent or undergo electrophilic substitution to provide rearrangement-like products.

Photoinduced electron transfer bond cleavage requires that the lowest unoccupied molecular orbital be σ^* in nature and localized on the sulfonium moiety. The rate of PET was determined by the thermodynamics and the distance between D and A moieties.

REFERENCES

1. Crivello, J. V. *Adv. Polym. Sci.* **1984**, *62*, 1.
2. Dektar, J. L.; Hacker, N. P. *J. Am. Chem. Soc.* **1990**, *112*, 6004, and references cited therein.
3. Saeva, F. D. In *Topics in Current Chemistry*. Springer-Verlag: New York, 1990; Vol. 159, p 59, and references cited therein.
4. Knibbe, H.; Rehm, D.; Weller, A. *Ber. Bunsenges. Phys. Chem.* **1969**, *73*, 839.
5. Hine, J. In *Physical Organic Chemistry*. McGraw-Hill: Pennsylvania, 1962, p 87.
6. Perkins, C. W.; Martin, J. C.; Arduengo, A. J.; Alegria, A.; Kochi, J. K. *J. Am. Chem. Soc.* **1980**, *102*, 7753.
7. Anklam, E.; Margaretha, P. *Res. Chem. Intermediates* **1989**, *11*, 127.
8. Saeva, F. D.; Morgan, B. P. *J. Am. Chem. Soc.* **1984**, *106*, 4121.
9. Saeva, F. D.; Morgan, B. P.; Luss, H. R. *J. Org. Chem.* **1985**, *50*, 4360.
10. Dewar, M. J. S.; Zoebisch, E. G.; Healy, E. F.; Steward, J. J. P. *J. Am. Chem. Soc.* **1985**, *107*, 3902.

11. Breslin, D. T.; Saeva, F. D. *J. Org. Chem.* **1988,** *53,* 713.
12. Saeva, F. D.; Breslin, D. T. *J. Org. Chem.* **1989,** *54,* 712.
13. Saeva, F. D.; Breslin, D. T.; Martic, P. A. *J. Am. Chem. Soc.* **1989,** *111,* 1328.
14. Saeva, F. D.; Breslin, F. D.; Luss, H. R. *J. Am. Chem. Soc.* **1991,** *113,* 5333.
15. Burgdorff, C.; Kircher, T.; Lohmannsroben, H. G. *Spectrochim. Acta Part A* **1988,** *44,* 1137.
16. Saeva, F. D.; Martic, P. A.; Garcia, E. *J. Photochem. Photobiol.* **1994,** *76.*
17. Crivello, J. V. In *UV Curing: Science and Technology*; Pappas, S. P., Ed.; Technology Marketing Corporation: Stanford, CT, 1978, p 23.
18. Crivello, J. V. *Polym. Eng. Sci.* **1983,** *23,* 953.
19. Crivello, J. V. *Adv. Polym. Sci.* **1984,** *62,* 1.
20. Pappas, S. P. *Radiat. Curing* **1981,** *8,* 28.
21. Pappas, S. P. *Prog. Org. Coatings* **1985,** *13,* 35.
22. Pappas, S. P. *J. Imaging Tech.* **1985,** *11,* 146.
23. Pappas, S. P.; Jilek, J. H. *Photogr. Sci. Eng.* **1979,** *23,* 140.
24. Pappas, S. P.; Gatechair, L. R.; Pappas, B. C. *J. Photochem.* **1981,** *17,* 120.
25. Pappas, S. P.; Gatechair, L. R.; Jilek, J. H. *J. Polym. Sci. Polym. Chem. Ed.* **1984,** *22,* 77.
26. Pappas, S. P.; Pappas, B. C.; Gatechair, L. R.; Schnabel, W. J. *J. Polym. Sci. Polym. Chem. Ed.* **1984,** *22,* 69.
27. Hacker, N. P.; Leff, D. V.; Dektar, J. L. *J. Org. Chem.* **1991,** *56,* 2280.
28. Dektar, J. L.; Hacker, N. P. *J. Org. Chem.* **1991,** *56,* 1838.
29. Dektar, J. L.; Hacker, N. P. *J. Am. Chem. Soc.* **1990,** *112,* 6004.
30. Hacker. N. P.; Dektar, J. L. *Polym. Mater. Sci. Eng.* **1989,** *61,* 76.
31. Dektar, J. L.; Hacker, N. P. *J. Org. Chem.* **1990,** *55,* 639.
32. Dektar, J. L.; Hacker, N. P. *J. Chem. Soc. Chem. Commun.* **1987,** 1591.
33. Dektar, J. L.; Hacker, N. P. *J. Org. Chem.* **1988,** *53,* 1833.
34. He, X.; Huang, W.; Reiser, A. *J. Org. Chem.* **1992,** *57,* 759.

DYNAMICS FOR THE FORMATION AND DECAY OF RADICAL ION PAIRS FORMED THROUGH EXCITED-STATE ELECTRON TRANSFER REACTIONS

Kevin S. Peters

Advances in Electron Transfer Chemistry
Volume 4, pages 27–52.

ISBN: 1-55938-506-5

1. INTRODUCTION

An important problem in bimolecular electron transfer reactions in solution between an electron donor molecule D and a photoactivated electron acceptor molecule A^* is determining under what conditions the electron is transferred at molecular contact to give contact radical ion pairs (CRIP) or at long range to give solvent-separated radical ion pairs (SSRIP). Furthermore, it is important to understand how the dynamics of the interconversion between CRIP, SSRIP, and free radical ions (FI) depends on the molecular structure of the reacting species as well as the nature of the solvent. The competition among these processes plays a fundamental role in governing the outcome of the photochemical reaction. For example, if the desired reaction is a 2+2 photocycloaddition, then it is advantageous to minimize the initial formation of the SSRIP. Conversely, if high yields of free radical ions are needed, then the production of CRIP should be avoided. Thus, to optimize the yields of photochemical reactions it is crucial that the parameters controlling the efficiencies for CRIP, SSRIP, and FI generation be understood.

For the past thirty years numerous studies, both experimental and theoretical, have sought to examine the parameters that control the efficiencies for radical ion pair formation. Through these efforts we now have a basic understanding of the important parameters that govern the efficiencies of these reactions, although for a given molecular system we still can not predict with certainty when a CRIP or a SSRIP will be formed. It is the aim of this chapter to detail the developments in both theory and experiment that address processes of radical ion pair generation and their ensuing dynamics. This chapter will not present a comprehensive review of the subject but rather focus on several key experimental studies and theoretical investigations that served to develop our present understanding of the nature of these processes.

2. BACKGROUND

The nature of the radical ion pair produced on the quenching of an excited singlet state of an acceptor molecule A^* by the transfer of an electron from a donor molecule D was first addressed in the pioneering studies of Weller and his co-workers.[1] In their investigation of anthracene fluorescence quenching by diethylanaline in nonpolar solvents, they observed the appearance of a new structureless emission to the red of the original emission from anthracene.[2] They attributed the new emission to the formation of an exciplex, which, however, has been alternately described as a CRIP composed of a radical cation and a radical anion in direct contact. The direct contact between these molecular species results in a large electronic coupling leading to a high probability for light emission.[2] When the same fluorescence-quenching studies were carried out in polar solvents, no new light-emitting species were observed. This observation led to the proposal that in polar solvents, fluorescence quenching by electron transfer produces a radical anion and a radical cation that are partially solvated, a SSRIP. In a SSRIP, the two radical ions are separated by at least one solvent molecule, resulting in a small electronic coupling between the two species and thereby leading to a negligibly small probability for light emission.

The support for the proposal that SSRIP are formed on fluorescence quenching in polar solvents came from the examination of the anthracene fluorescence intensity as a function of diethylanaline concentration.[2] The concentration dependence of the fluorescence intensity is expressed as

$$I(c)/I(0) = \exp\{-V_D[I(c)/I(0)]^{1/2}c_D]/(1 + k\tau_0 c_D) \quad (1)$$

where c_D is the concentration of diethylanaline, τ_0 the lifetime of the anthracene, $I(0)$ the fluorescence intensity in the absence of diethylanaline, $I(c)$ the fluorescence intensity in the presence of diethylanaline at concentration c_D, k the rate constant for the quenching, and V_D the molar volume of diffusion, given by

$$V_D = 4\pi N'(\gamma a)^2[\tau_0(D_A + D_D)]^{1/2} \quad (2)$$

N' is the number of molecules per millimole, $D_A + D_D$ is the sum of the diffusion coefficients for anthracene and diethylanaline, and γa is the effective encounter distance for electron transfer where γ is the probability for electron transfer at reaction distance a. For the fluorescence quenching of anthracene by diethylanaline in acetonitrile, the encounter distance is $a = 7$ Å, supporting the proposal that SSRIP are formed on electron transfer in acetonitrile. The free energy change for this reaction is estimated to be $\Delta G = -14$ kcal/mole based on the Weller equation[2] for the free energy difference between $^1A^* + D$ and SSRIP,

$$\Delta G = E(D/D^+) - E(A^-/A) - \Delta E_{00} - e_0^2/\varepsilon a \quad (3)$$

where $E(D/D^+)$ is the oxidation potential for diethylanaline, $E(A^-/A)$ is the reduction potential for anthracene, ΔE_{00} is the energy of the zero–zero electronic transition of anthracene, and $e_0{}^2/\varepsilon a$ is the coulombic energy for the radical ions at a separation a in a medium with dielectric constant ε.

Following these initial studies, Rehm and Weller proceeded to examine the kinetics for the fluorescence quenching of over 60 donor–acceptor pairs in acetonitrile; the ΔG for these reactions span the range of +5 kcal/mole to –60 kcal/mole.[3] The kinetics for fluorescence quenching increased from 5×10^6 M^{-1} s^{-1} to 2×10^{10} M^{-1} s^{-1} for the reaction free energy change from +5 kcal/mole to –10 kcal/mole. From –10 kcal/mole to –60 kcal/mole, the kinetics for fluorescence quenching remained constant at the diffusion limit, 2×10^{10} M^{-1} s^{-1}. The presumed mechanism for the fluorescence quenching for each of the donor–acceptor pairs was that of outer-sphere electron transfer producing SSRIP. Since the free energy dependence of the rate of electron transfer deviated in the highly exothermic regime ($\Delta G < -25$ kcal/mole) from the predictions of Marcus theory[4], Rehm and Weller proposed that low-lying electronic excited states of the radical ions may intervene in the quenching process for highly exothermic reactions. This suggestion, however, has yet to be validated. Importantly, for the range in free energies +5 kcal/mole to –25 kcal/mole in acetonitrile, it was proposed that fluorescence quenching by electron transfer leads directly to SSRIP.

Based on fluorescence quenching experiments as well as flash photolysis studies of the quenching of the first excited singlet state of pyrene by various aromatic and aliphatic amines, Weller proposed the following reaction scheme to serve as a model of the reaction pathways for the formation and annihilation of radical ion pairs; Scheme 1 has been modified so as not to include triplet pathways and diffusional recombination of free radical ions.[5,6]

In this model the competition between the formation of SSRIP and that of CRIP is governed by the relative magnitudes of the rate k_{let} of

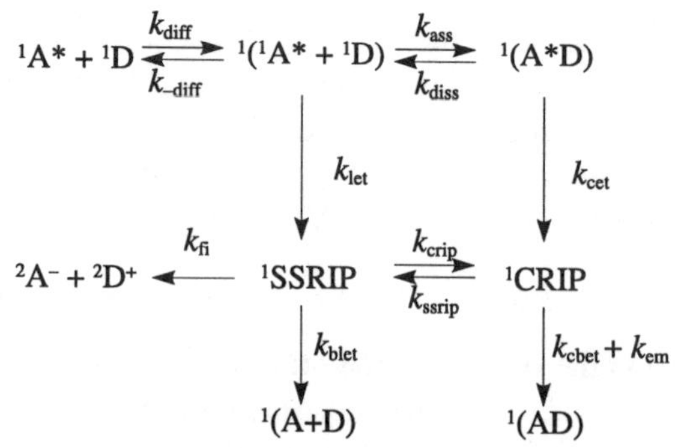

$$
\begin{array}{ccccc}
{}^1A^* + {}^1D & \underset{k_{-diff}}{\overset{k_{diff}}{\rightleftharpoons}} & {}^1({}^1A^* + {}^1D) & \underset{k_{diss}}{\overset{k_{ass}}{\rightleftharpoons}} & {}^1(A^*D) \\
 & & \downarrow k_{let} & & \downarrow k_{cet} \\
{}^2A^- + {}^2D^+ & \overset{k_{fi}}{\longleftarrow} & {}^1SSRIP & \underset{k_{ssrip}}{\overset{k_{crip}}{\rightleftharpoons}} & {}^1CRIP \\
 & & \downarrow k_{blet} & & \downarrow k_{cbet} + k_{em} \\
 & & {}^1(A+D) & & {}^1(AD)
\end{array}
$$

Scheme 1.

long-range electron transfer and the rate of collapse of the geminate pair to the contact pair, $^1(^1A^* + ^1D) \rightarrow {}^1(A^*D)$ followed by contact electron transfer, k_{cet}. The theoretical formulations for k_{let} and k_{cet} are presented in the following section. In the fluorescence-quenching experiments, Weller envisioned that for the energy regime –20 kcal/mole < ΔG < 5 kcal/mole, the geminate pair $^1(^1A^* + ^1D)$ decays principally to ^{1}SSRIP. However, in these latter studies, Weller has suggested[5] that the lack of observation of the inverted region for large negative free energy changes ($\Delta G < -20$ kcal/mole) results in the formation of the exciplex (CRIP), as k_{ass} followed by k_{cet} ($\geq 10^{12}$ s^{-1}) is greater than k_{let} in this free energy regime. However, this proposal is not supported by recent developments in theories of electron transfer.

The rate constant k_{ssrip} for CRIP separation to SSRIP was estimated based on measurements of exciplex emission lifetimes.[5] The values for k_{ssrip} range from 5×10^8 s^{-1} in acetonitrile to $< 10^6$ s^{-1} in tetrahyrdofuran. Based on these experiments, the solvent dependence for the rate of CRIP separation is described by the following empirical equation:

$$k_{ssrip} = 2.3 \times 10^9\, \eta^{-1} \exp\,[-e^2_0/\varepsilon k\mathrm{T}\,(1/d - 1/a)] \qquad (4)$$

where η is the viscosity of the solvent, ε the dielectric of the medium, k the Boltzmann constant, and a and d the initial and final separations of radical ion pairs. Although there is no direct measurement of k_{crip}, this value may be estimated from the equilibrium constant between CRIP and SSRIP. Weller developed a method for calculating the free energy change for radical ion pair interconversion.[7] The formalism is based upon the Kirkwood–Onsager model for the solvation of a dipole and the Born equation for the solvation of an ion. The free energy change (in eV) as a function of the dielectric constant for the medium in the conversion of a contact radical ion pair to a solvent separated radical ion pair is

$$\Delta G = (\mu^2/\rho^3)\,[(\varepsilon - 1)/(2\varepsilon + 1) - 0.19] + 2.6\varepsilon - 0.51 \qquad (5)$$

where μ is the dipole moment of the CRIP, ρ the radius of the solvation cavity of the CRIP, and ε is the static dielectric constant of the solvent. This result assumes that the mixing of the radical ion pair states with the ground state reactants and the excited singlet state is negligible. For radical ion pairs in acetonitrile, the free energy change associated with the conversion of a CRIP into a SSRIP is approximately $\Delta G = -5$ kcal/mole. With a value of 5×10^8 s^{-1} for k_{ssrip}, an estimate of the rate of collapse of a SSRIP into a CRIP in acetonitrile is $k_{crip} = 1.2 \times 10^5$ s^{-1}.

3. THEORETICAL FORMULATIONS FOR ELECTRON TRANSFER

To develop an understanding of the parameters that control the efficiency of electron transfer in geminate, $^1(^1A^*+^1D)$, and contact, $^1(A^*D)$, donor–acceptor pairs, one may gain insight by examining the theoretical formulations for the rates of electron transfer. In the geminate pair, where a solvent molecule separates the reacting species, the electronic coupling between A^* and D is small. For the limit of weak coupling, nonadiabatic theories for electron transfer are applicable. The rate of nonadiabatic electron transfer, based upon Marcus's original formulation of electron transfer theory, is expressed as[8,9]

$$k_{\mathrm{let}} = (4\pi^3/h^2\lambda_s k_b T)^{1/2} V^2 \sum_{j=0}^{\infty} F_j \exp[-(\Delta G + hv + \lambda_s)^2/4\lambda_s k_b T] \quad (6)$$

where $F_j = e^{-S}S^j/j!$ and $S = \lambda_v/hv$. The vibration v corresponds to a single active mode inducing the electron transfer process. The energy associated with the reorganization of the nuclei that accompany the electron transfer is λ_v while the energy associated with the change in the solvent structure is λ_s. The free energy change for the electron transfer is ΔG. The electronic coupling between A^* and D is V.

When the donor and acceptor are in direct molecular contact, $^1(A^*D)$, coupling between the reacting species will be large and thus it is necessary to formulate the rate k_{cet} of contact electron transfer within an adiabatic theory for electron transfer. One such formulation is that of Jortner and Bixon,[10] where k_{cet} is expressed as

$$k_{\mathrm{cet}} = \sum_j k_{\mathrm{NA}}(j)/(1 + H_{\mathrm{A}}(j)) \quad (7)$$

where k_{NA} is the nonadiabatic rate constant defined in Eq. 6. The term H_{A} is the adiabatic parameter defined as

$$H_{\mathrm{A}}(j) = 4\pi F_j V^2 \tau_{\mathrm{L}}/\hbar\lambda_s \quad (8)$$

The longitudinal dielectric relaxation time τ_{L} is obtained through the relationship $\tau_{\mathrm{L}} = (\varepsilon_\infty/\varepsilon_s)\tau_{\mathrm{D}}$ where ε_∞ and ε_s are the high frequency and static dielectric constants for the solvent and τ_{D} is the Debye solvent relaxation time.

The applicability of Jortner–Bixon theory of adiabatic electron transfer to the kinetics of charge separation on formation of the contact pair, k_{cet}, has not been examined from the vantage point of experiment. There is no experimental evidence for the existence of an inverted region for k_{cet} similar to that found in nonadiabatic electron transfer. Indeed, the rate k_{cet} of charge separation may be a function only of the kinetics for solvent reorganization, on the order of 10^{12} s^{-1}, to accommodate the separation of charge.[11] With this caveat, the ensuing discussion of the kinetics of exciplex creation on contact pair formation will be treated within the Jortner–Bixon formalism.

The efficiencies for the creation of CRIP and SSRIP are critically dependent on the relative magnitudes of k_{let} and k_{cet}, as well as k_{ass}. Of particular interest to the present discussion is how these two rate constants depend on ΔG for the electron transfer reaction. Before the functional dependence of k_{let} and k_{cet} on ΔG can be determined employing Eqs. (6) and (7), the parameters V, λ_v, and λ_s must be estimated for the geminate and contact pairs. These parameters are functions of the nature of the reacting pair, contact or geminate, as well as of the electronic properties of the molecule and the solvent. The electronic coupling matrix element V depends exponentially upon the distance for separation of the reacting pair.[12] The most common expression for V as a function of the separation distance R is

$$V = V_0 \exp\left[-\beta(R - R_0)/2\right] \tag{9}$$

where V_0 is the electronic coupling at the van der Waals distance R_0 and β the distance dependence of the scaling parameter. For intermolecular electron transfer, β is on the order of 2Å^{-1}. The solvent reorganization energy λ_s is estimated though the relationship[13]

$$\lambda_s = e^2[(2r_D)^{-1} + (2r_A)^{-1} - (R_{DA})^{-1}](\varepsilon_s^{-1} + \varepsilon_\infty^{-1}) \tag{10}$$

where r_A and r_D are the radius of the acceptor and donor, R_{DA} the intermolecular separation of the reacting species, and ε_s and ε_∞ the static and high-frequency dielectric constants of the solvent. Finally, the vibrational reorganization energy λ_v to the first approximation is independent of the nature of the reacting pair (contact or geminate).

For intermolecular electron transfer reactions only a limited number of studies have sought to obtain the parameters used in the calculation of the rates of electron transfer. The most extensive study to date is that of Gould and Farid and their co-workers.[14] They have measured the rates of electron backtransfer within CRIP and SSRIP for tetracyanoanthracene radical anion and substituted benzene radical cations in acetonitrile. In the CRIP the electronic coupling is $V = 1000\ \text{cm}^{-1}$ and the solvent reorganization energy is $\lambda_s = 0.55$ eV, while in the SSRIP the electronic coupling decreases to $V = 10.8\ \text{cm}^{-1}$ and the solvent reorganization energy increases to $\lambda_s = 1.72$ eV. The vibrational reorganization energy is $\lambda_v = 0.2$ eV. At this time it is not clear how sensitive these various parameters are to the molecular and electronic structure of the reacting species. Also, there has been some discussion[15] as to whether these parameters, the solvent reorganization energy, particularly derived from charge recombination reactions are applicable to charge separation reactions, which is the focus of the present discussion. More studies of both charge separation and charge recombination reactions are needed to address these issues fully.

Employing the above parameters in Eqs. (6) and (7) the rate constants k_{let} and k_{cet} as a function of ΔG are shown in Figure 1. The rate k_{cet} of

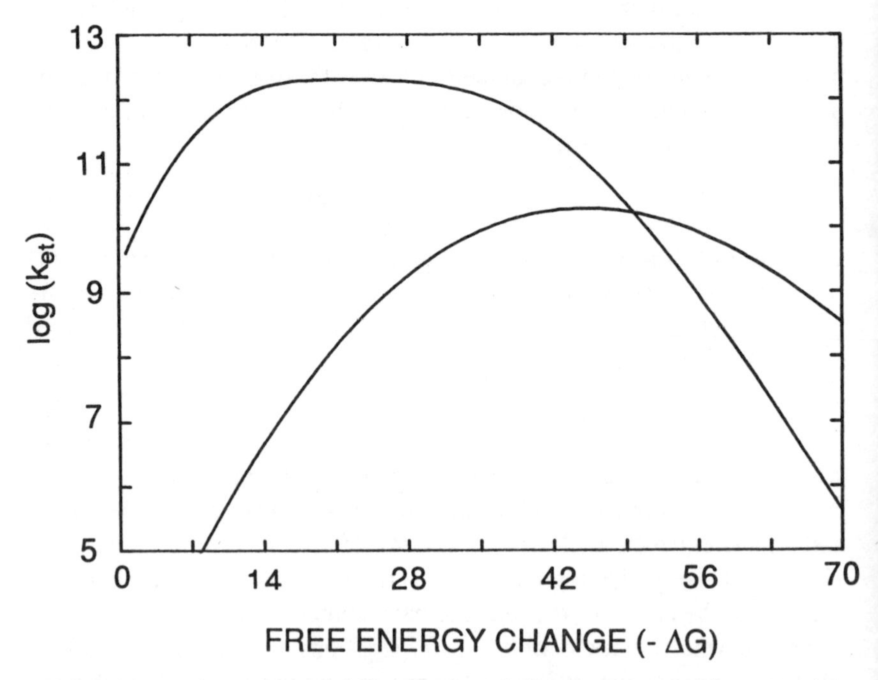

Figure 1. Upper curve: Adiabatic rate of electron transfer from Eq. (7), where $V = 1000\ \text{cm}^{-1}$, $\lambda_s = 0.55$ eV, $\lambda_v = 0.2$ eV, $v = 1400\ \text{cm}^{-1}$, and $\tau_l = 7 \times 10^{-13}\ \text{s}^{-1}$. Lower curve: Nonadiabatic rate of electron transfer from Eq. (6), where $V = 10.8\ \text{cm}^{-1}$, $\lambda_s = 1.72$ eV, $\lambda_v = 0.2$ eV, and $v = 1400\ \text{cm}^{-1}$.

electron transfer at contact increases from 1×10^{10} to 3×10^{12} s^{-1} for –14 kcal/mole $\leq \Delta G \leq 0$ kcal/mole and then decreases from 10^{12} to 5×10^5 s^{-1} for –70 kcal/mole $\leq \Delta G \leq$ –35 kcal/mole. Conversely, the rate of electron transfer within the geminate pair increases from 1×10^3 to 3×10^{10} s^{-1} for –42 kcal/mole $\leq \Delta G \leq 0$ kcal/mole and then decreases from 3×10^{10} to 5×10^8 s^{-1} for –70 kcal/mole $\leq \Delta G \leq$ –42 kcal/mole.

These calculations reveal that the rate of electron transfer within the contact pair is greater than within the geminate pair for –50 kcal/mole $\leq \Delta G \leq 0$ kcal/mole. However the yield of CRIP and SSRIP also depends on k_{ass}, the rate for collapse of the geminate pair to the contact pair, as well as the rate k_{-diff} of diffusional separation of the geminate pair. Over the energy range of 0 to –40 kcal/mole, the approximate yield Φ of CRIP and SSRIP creation following the formation of the geminate pair is given by

$$\Phi_{CRIP} = k_{ass}/(k_{ass} + k_{let} + k_{-diff}) \tag{11}$$

and

$$\Phi_{SSRIP} = k_{let}/(k_{ass} + k_{let} + k_{-diff}) \tag{12}$$

which also assumes that $k_{cet} >> k_{diss}$. The yield R of CRIP formation relative to SSRIP formation is then

$$R = \Phi_{CRIP}/\Phi_{SSRIP} = k_{ass}/k_{let} \tag{13}$$

Although the values of k_{let} and k_{cet} can be determined as functions of ΔG, virtually nothing is known experimentally about the rate k_{ass}; its value is probably on the order of 10^{10} s^{-1}. With this assumption for k_{ass}, the yield of CRIP will exceed that of the SSRIP for the range –35 kcal/mole $\leq \Delta G \leq 0$. For $\Delta G \leq$ –35 kcal/mole, the yield of SSRIP will be greater than that of the CRIP.

In ascertaining whether the electron transfer proceeds in the contact pair or the geminate pair, it is important to consider the concentrations of the reacting species. For example, if the concentration of the electron donor is on the order of 1 M, then a major fraction of the acceptor molecules have a donor molecule within the first solvent shell. If ΔG for the reaction is –40 kcal/mole, then the acceptors will be quenched while at molecular contact with the donors. On the other hand, when the same reaction is carried out at low concentrations of the donor so that the initial concentration of the contact pair is low, then in this energy regime a large fraction of the electron transfer reactions may proceed within the geminate pair, assuming $k_{let} > k_{ass}$. Thus to address the question of whether the electron transfer proceeds within the contact or geminate pair it is necessary to specify the initial conditions for the experiment.

4. KINETIC STUDIES OF RADICAL ION PAIR FORMATION AND DECAY

The following section details a number of investigations that examine the kinetics of the formation and decay of both CRIP and SSRIP. The discussion will initially focus on the nature of the radical ion pair formed on diffusional quenching of a photoexcited molecule. The kinetic behavior of electron backtransfer within the CRIP and SSRIP is then presented and then followed by a survey of the dynamics for radical ion pair diffusional separation. Finally, the kinetics for intersystem crossing within the radical ion pair are discussed.

4.1. Contributions of k_{let} and k_{cet} to the Quenching of A^*

Pyrene-Pyromellitic Dianhydride and Pyrene-Tetracyanoethylene

In a series of picosecond kinetic studies, Mataga and co-workers have examined the kinetics of charge separation and charge recombination within a variety of reactants.[16–21] Their particular emphasis has been on establishing the relationship between the rates of electron transfer and the free energy change for the reactions. Also, they have examined the kinetics of translational diffusion between radical ion pair forms. Two systems which have been thoroughly examined[17,18] are the electron transfer reactions between pyrene (PY) and pyromellitic dianhydride (PMDA) as well as pyrene and tetracyanoethylene (TCNE).

In the ground state, PY forms a charge-transfer complex with both PMDA and TCNE.[17] Both of these charge-transfer complexes absorb at 530 nm in acetonitrile. Excitation of the PY/PMDA charge-transfer complex results in the direct transfer of an electron from PY to PMDA to form the radical ion pair $PY^{+}/PMDA^{-}$. Similarly, excitation of the PY/TCNE charge-transfer complex results in the formation of $PY^{+}/TCNE^{-}$. It is assumed that the radical ion pairs are created in the contact form (CRIP), as the ground-state charge-transfer complex exists only when the two molecules are in direct molecular contact.

The time evolution of the CRIP is monitored through the absorption change of the PY^{+} chromophore at 455 nm using femtosecond–picosecond absorption spectroscopy.[17] The lifetimes for the PY/PMDA and PY/TCNE CRIP in acetonitrile are 10 ps and 1 ps, respectively. The decay of the CRIP proceeds by rapid electron backtransfer to form ground-state reactants k_{cbet}. There is no evidence for CRIP diffusional separation to the SSRIP or FI forms. Importantly, no emission was observed from either of the two CRIP. In prior experiments the lack of exciplex emission upon diffusional quenching has been viewed as evidence for SSRIP formation.[2] Clearly, CRIP may be formed and decay by rapid electron backtransfer k_{cbet}, resulting in no observable exciplex emission.

The diffusional quenching of the first excited singlet state of pyrene, PY^*, by PMDA and TCNE was also examined.[16–18] Under these conditions the kinetic behavior of the resulting radical ion pairs was significantly different from that of the CRIP formed upon irradiation of the ground-state charge-transfer complex. The lifetime of the PY/PMDA radical ion pair is 80 ps, and the lifetime of the PY/TCNE radical ion pair is 285 ps in acetonitrile. Both radical ion pairs decayed by two processes, charge recombination to form ground state reactants k_{blet} and diffusional separation to form long-lived (> 100 ns) radical ion pairs k_{fi}. Based upon the kinetic characteristic of the various radical ion pairs, Mataga proposed[16] that diffusional quenching leads to the formation of SSRIP for each of the reacting pairs. However in both of these experimental studies, the absolute yield of the SSRIP was not established. If CRIP were formed on diffusional quenching, its presence would not be manifested in the kinetic data, given the exceedingly short lifetime of the CRIP as established in the kinetic studies of the charge-transfer complexes. Thus there is clear evidence for formation of the SSRIP although the role of the CRIP in the diffusional quenching remains to be defined for these molecular systems.

These experiments are in accord with the theoretical predictions contained in Figure 1. Based upon Eq. (3), the free energy changes associated with the quenching of PY^* by PMDA and TCNE are $\Delta G = -32$ kcal/mole and $\Delta G = -50$ kcal/mole.[16] At –50 kcal/mole, the rate of electron transfer in the geminate pair leading to SSRIP is approximately equal to the rate of electron transfer in the contact pair leading to CRIP, so that formation of the SSRIP is kinetically feasible. At $\Delta G = -32$ kcal/mole, the rate of electron transfer within the contact form greatly exceeds the rate within the geminate form. However, the rate of collapse of the geminate pair to the contact pair k_{ass} plays a pivotal role in establishing the relative yields of CRIP and SSRIP; that SSRIP are observed for the quenching of PY^* by PMDA is consistent with the present theoretical predictions that CRIP and SSRIP should be formed in approximately equal amounts, based on Eq. (13).

In summary, these studies reveal that in acetonitrile long range electron transfer to form the SSRIP is a kinetically important process in bimolecular electron transfer reactions. The partitioning between CRIP and SSRIP formation has yet to be delineated.

Exciplex Emission Studies in Acetonitrile

In Wellers's fluorescence quenching studies where the solvent was acetonitrile, no exciplex emission was observed.[3] Based on this observation, the mechanism of electron transfer was assumed to proceed by an outer-sphere mechanism giving rise to SSRIP. This proposal was further supported by the determination of the effective encounter distance of 7 Å between the fluorescer and the quencher.[2] The assumption has been that all excited-state electron transfer reactions in acetonitrile produced SSRIP.

The first report of exciplex emission in fluorescence quenching experiments in acetonitrile was by Kikuchi and co-workers.[22] They investigated the fluorescence quenching of anthracenecarbonitriles by a series of aromatic quenchers. When ΔG for the electron transfer ranged from +4.6 to –6.4 kcal/mole, exciplex emission was observed. For electron transfer reactions with $\Delta G \leq -10.8$ kcal/mole, no emission was observed. To account for this behavior, these workers proposed that in the range -6.4 kcal/mole $\leq \Delta G \leq 4.6$ kcal/mole the electron transfer occurs when the two molecules are in direct molecular contact, and when $\Delta G \leq -10.8$ kcal/mole the electron transfer occurs when the two molecules are in the geminate pair. It is noted that in these studies, the quantum yield for exciplex emission was not measured, and thus it was not possible to determine the yields for the formation of the various radical ion pairs.

Gould and Farid presented a more extensive study of exciplex emission in acetonitrile, where the quantum efficiencies of the exciplex formation were determined.[23,24] Their analysis was based on the following kinetic Scheme 2. The aim of their investigation was to determine the efficiency α of the formation of the contact pair $^1(A^*D)$, producing the exciplex ^{1}CRIP. The yield of ^{1}CRIP was determined by its fluorescence quantum yield, Φ_f, where

$$\Phi_f = \alpha \, k_{em} \, \tau \tag{14}$$

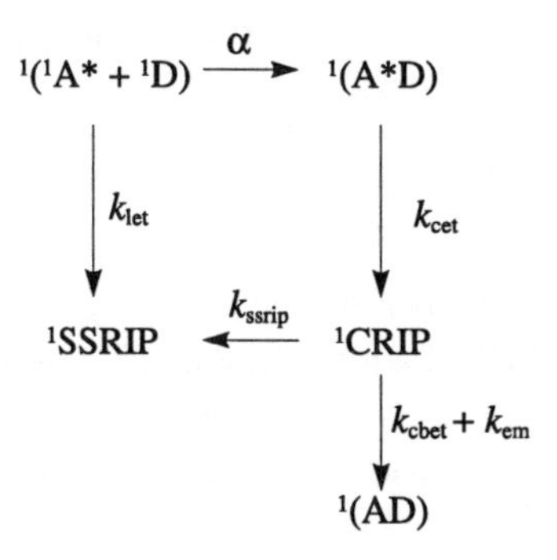

Scheme 2.

The terms k_{em} and τ are the radiative rate constant and the lifetime of the ^{1}CRIP. The lifetime is controlled by three relaxation processes: the rate k_{ssrip} of CRIP separation to the SSRIP, the rate k_{cbet} of electron backtransfer within the CRIP and the rate k_{em} of fluorescence. The lifetime is $\tau = (k_{ssrip} + k_{cbet} + k_{em})$. By experimentally determining Φ_f and τ, and employing k_f which can be measured in nonpolar solvents where $\alpha = 1$, the efficiency of exciplex formation is determined.

Gould and Farid examined the quenching of dicyanoanthracene by a series of alkylbenzenes where ΔG for the electron transfer reaction varied from –6.7 to –15.0 kcal/mole.[23] For the range -8.5 kcal/mole $\leq \Delta G \leq -6.7$ kcal/mole, the yield of exciplex formation approached unity; the rate decreased to 0.25 for $\Delta G = -11.7$ kcal/mole and then become negligibly small for $\Delta G = -15.0$ kcal/mole.

The importance of these studies is that they clearly demonstrate that in acetonitrile both short- and long-range electron transfer reactions can occur. More important, however, is that their observations are not in accord with the prediction of the free energy dependence for the kinetics of adiabatic and nonadiabatic electron transfer shown in Figure 1. For electron transfer with $\Delta G = -15$ kcal/mole, the rate of electron transfer within the geminate is calculated to be $k_{let} = 5 \times 10^7$ s^{-1}. Thus, based on Eq. (13), the expected yield of the CRIP should be large, contrary to observation. The source of this discrepancy between theoretical prediction and experimental observation may be that the parameters used in the calculation of k_{let} in Eq. (6) and k_{cet} in Eq. (7) may not be appropriate for charge separation in the molecular systems of Gould and Farid. Again it is noted that the parameters used in the calculations shown in Figure 1 were derived from charge recombination within the dicyanoanthracene–alkylbenzene radical ion pairs.

Stilbene–Electron Deficient Alkenes

The nature of the radical ion pair produced on the quenching of the first excited singlet state of *trans*-stilbene (S_1) by fumaronitrile (FN) in acetonitrile has recently been addressed by Peters and co-workers.[24–30] The free energy change for this reaction is estimated to be –17.5 kcal/mole.[27] In this study, the kinetics of the quenching of the first excited singlet state were examined and the distance parameter for electron transfer was deduced. In order to ascertain whether CRIP or SSRIP are formed on electron transfer from S_1 to FN, the kinetic properties of the CRIP and SSRIP were obtained and compared to the radical ion pair dynamics produced upon diffusional quenching.[29]

The dynamics of the quenching of S_1 in the presence of FN were examined.[28] The rate constant for the quenching of S_1 was found to be time dependent. The origin of the time-dependent rate constant for the quenching of S_1 by FN is a result of the nature of the original distribution of FN about S_1. Prior to the excitation of *trans*-stilbene (TS), there is an equilibrium distribution of FN about TS so that at high concentrations some of the TS have a molecule of FN within the first solvent shell. Upon excitation, those S_1 with a molecule of FN in the vicinity are rapidly quenched and the rate of the reaction is limited by k_{et}. However, for those S_1 which do not have a molecule of FN in the vicinity, the quenching of S_1 depends on the diffusion of FN and is limited by k_d. A theory for solution-phase kinetics when $k_{et} \geq k_d$ was developed by Collins and Kimball.[31] Solving Fick's second law of diffusion with the boundary condition $c(R) = (D/\kappa)\,(\delta c/\delta r)_R$ yields

$$k_{CK}(t) = [k_{et}^{-1} + k_d^{-1}]^{-1}[1 + (k_{et}/k_d)\exp(y^2)\,\mathrm{erfc}(y)] \tag{15}$$

where

$$\mathrm{erfc}(y) = (2\pi)^{-1/2}\int_y^\infty \exp(-u^2)\,du \tag{16}$$

and

$$y = \{(Dt)^{1/2}/R\}\,\{1 + (k_{et}/k_d)\} \tag{17}$$

The parameter $D = D_A + D_D$ is the sum of the diffusion coefficients for the acceptor and donor molecules, which can be determined by electrochemical measurements. R is the distance at which the electron transfer reaction occurs. Since k_d can be described as

$$k_d = 4\pi R N_A(D_D + D_A) \tag{18}$$

where N_A is Avogadro's number, the only unknown parameters in the above equations are k_{et} and R, since D_D and D_A can be determined experimentally.[28]

The Collins and Kimball formalism allows for the determination of k_{et} in the absence of diffusion, so that the intrinsic rate of electron transfer can be measured. For the quenching of S_1 by FN, $k_{et} = (3.0 \pm 1.0) \times 10^{11}$ s^{-1}. This is the first direct measurement of the rate of an intermolecular electron transfer process that is greater than the diffusion-controlled rate.

To explore further the dependence of the rate of electron transfer on ΔG, the quenching of the first excited singlet state of TS by acrylonitrile (AN) and tetracyanoethylene (TCE) as well as the first excited singlet state of 4,4′-dimethyl-*trans*-stilbene (DMES) by AN, FN, and TCE were examined (see Table 1).[28] The most intriguing aspect of this series of experiments is the relationship between the rate of electron transfer and the corresponding free energy change for the reaction. When the reaction is endoergic, + 7.1 kcal/mole for TS*/AN, the rate of electron transfer is less than the diffusion-controlled rate: $k_{et}(TS^*/AN) = 2.6 \times 10^9\ M^{-1}\ s^{-1}$. When the reaction becomes exoergic, –17.5 kcal/mole for TS*/FN, the rate constant increases to $k_{et}(TS^*/FN) = 3.0 \times 10^{11}\ M^{-1}\ s^{-1}$. Surprisingly, when the reaction becomes highly exoergic (–52.8 kcal/mole for TS*/TCE), the rate of electron transfer does not decrease as predicted by

Table 1. Summary of Calculated and Experimentally Observed Electron Transfer Reactions in Acetonitrile

Reactants	*Products*	ΔG^a *(kcal/mole)*	k_{et} *($M^{-1}\ s^{-1}$)*	*R (Å)*
TS*/AN	TS+/AN–	+ 7.1	$2.6 \pm 0.1 \times 10^9$	
DMES*/AN	DMES+/AN–	+ 4.6	$5.6 \pm 0.1 \times 10^9$	
TS*/FN	TS+/FN–	–17.5	$3.0 \pm 1.0 \times 10^{11}$	8.7
DMES*/FN	DMES+/AN–	–20.1	$4.0 \pm 0.6 \times 10^{11}$	10.5
TS*/TCE	TS+/TCE–	–52.8	$4.5 \pm 1.2 \times 10^{11}$	10.5
DMES/TCE	DMES+/TCE–	–55.3	$4.0 \pm 1.2 \times 10^{11}$	11.0

Note: [a]Free energy change based on Eq. (3).

Marcus electron transfer theory, but rather remains constant within the error of the experiment: $k_{et}(TS^*/TCE) = 4.5 \times 10^{11}\ M^{-1}\ s^{-1}$. However, as with the Rehm–Weller experiment for the highly exoergic reactions TS^*/TCE and DMES/TCE, the participation of the excited states of TS^+/TCE^- and $DMES^+/TCE^-$ in the electron transfer process can not be ruled out.[3]

The distance for the electron transfer between the first excited singlet state of TS and FN is $R = 8.7$ Å, and thus it would appear that the quenching of S_1 by FN leads to the creation of a SSRIP. However, the Collins and Kimball formalism for time-dependent diffusional quenching assumes that the reacting molecules can be represented as diffusing spheres and the bulk diffusion coefficients are parameters that are applicable to contact and geminate pairs.[31] Clearly it is important to

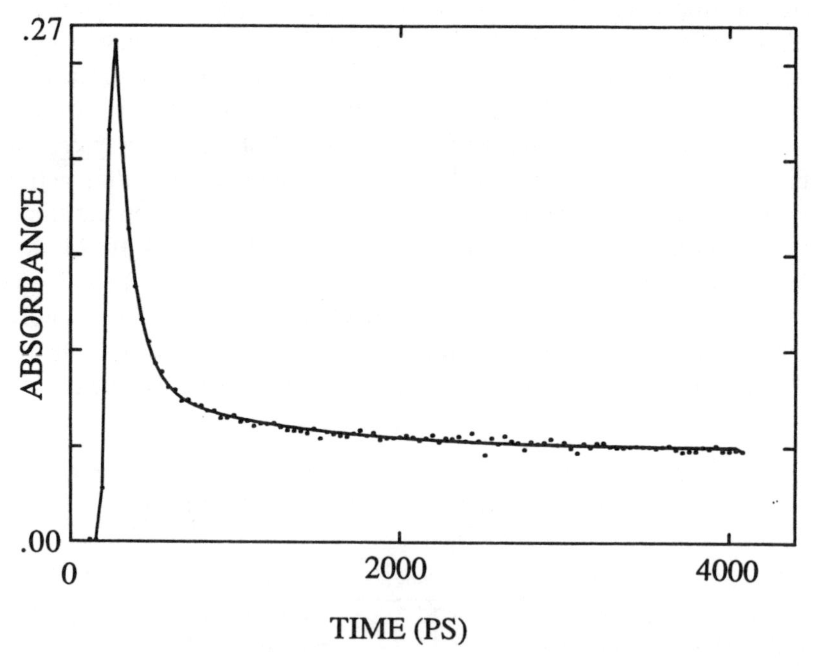

Figure 2. Dynamics of the *trans*-stilbene radical cation, monitoring at 480 nm following the 355-nm irradiation of a solution of 0.125 M FN, 0.025 M TS in acetonitrile. Points: experimental data. Solid curve: calculated kinetics based on Scheme 3 and the associated rate constants given in Table 2.

examine the kinetic characteristics of the radical ion pairs formed on the quenching of S_1 by FN and compare this behavior to the kinetic properties for the TS/FN CRIP and SSRIP.

In acetonitrile, TS combines with FN to form a ground-state charge-transfer complex, as evidenced by the appearance of a new absorption band at 355 nm. Irradiation of the charge-transfer band induces the transfer of an electron from TS to FN, forming the CRIP directly. The dynamics of the CRIP are monitored through the time evolution of the radical cation absorption of TS, TS^+, absorbing at 480 nm.[29] As shown in Figure 2, there is an initial rapid decay in TS^+ absorption during the first 300 ps followed by a slower decay during the subsequent 3 ns. The following kinetic scheme was used in analyzing the decay of the CRIP:

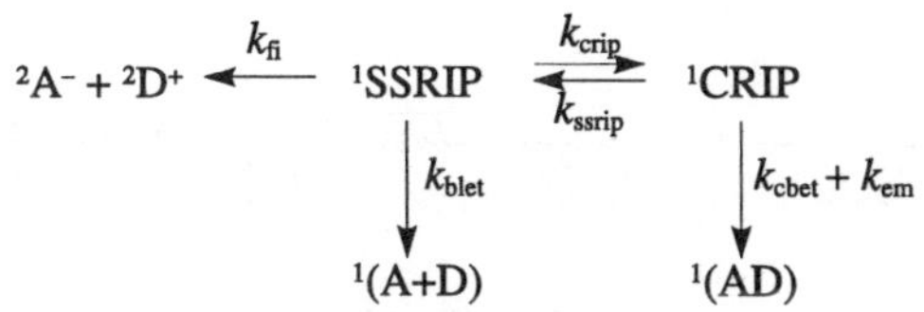

$$
\begin{array}{ccccc}
{}^2A^- + {}^2D^+ & \xleftarrow{k_{fi}} & {}^1SSRIP & \underset{k_{ssrip}}{\overset{k_{crip}}{\rightleftharpoons}} & {}^1CRIP \\
 & & \downarrow k_{blet} & & \downarrow k_{cbet} + k_{em} \\
 & & {}^1(A+D) & & {}^1(AD)
\end{array}
$$

Scheme 3.

The initial CRIP may decay by one of three pathways, and two of these decay processes lead to a decrease in the TS^+ absorbance. The processes leading to a decrease in TS^+ absorption are electron backtransfer within the CRIP (k_{cbet}) and exciplex emission (k_{em}). The third decay process, which does not lead to a decrease in TS^+ absorption, is radical ion pair separation to the SSRIP (k_{ssrip}). These kinetic processes are reflected in the time evolution of the TS^+ absorbance as a single decay. Thus to explain the 3-ns decay in TS^+, it is necessary to assume that the SSRIP undergoes three relaxation processes. Two processes, which will not directly effect the TS^+ absorbance, are collapse to the CRIP (k_{crip}) and separation to FI (k_{fi}). The only process within the SSRIP that causes a decrease in the TS^+ absorbance is long-range electron backtransfer (k_{blet}).

From the analysis of these and related experiments it was deduced that k_{blet} is significantly greater than k_{crip}.[30] The rate constants derived for the decay of the TS/FN radical ion pairs are given in Table 2.

The kinetics of the radical ion pair formed on the quenching of S_1 by FN are shown in Figure 3 and were analyzed within the following kinetic scheme:

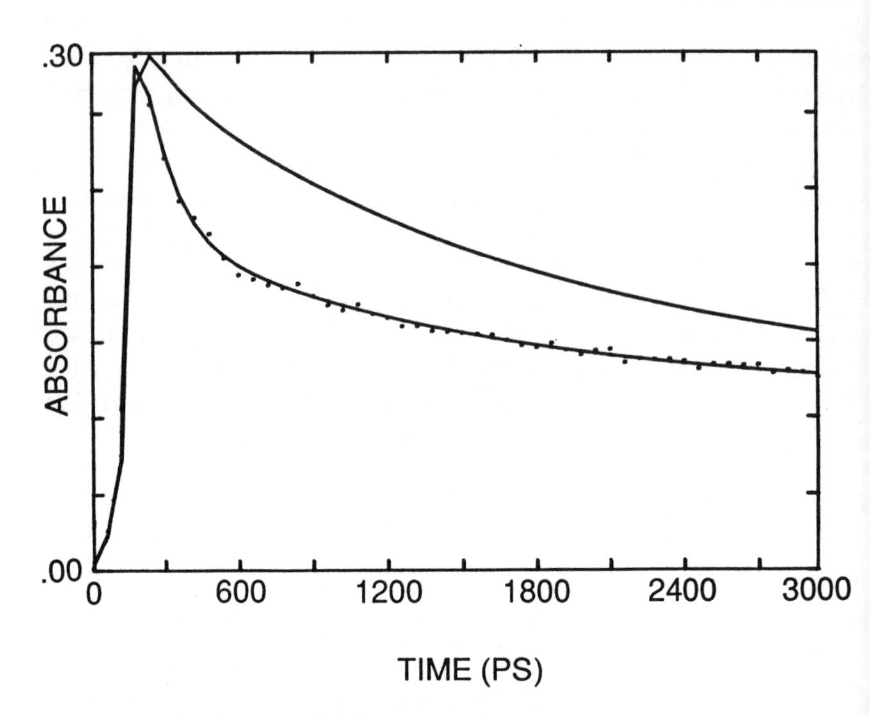

Figure 3. The dynamics of the *trans*-stilbene radical cation, monitoring at 480 nm, following the 266-nm excitation of 0.6 mM TS and 0.2 M FN in acetonitrile. Points: experimental data. Top solid curve: Calculated decay based on Scheme 4 with $k_{cet} = 0.0$ and $k_{let} = 2.63 \times 10^{10}\ s^{-1}$ and the rate coefficients listed in Table 2. Lower solid curve: Calculated decay based on Scheme 3 with $k_{cet} = 2.63 \times 10^{10}\ s^{-1}$, and $k_{let} = 0.0$, and the rate coefficients listed in Table 2.

$S_1 + FN$

k_{let} k_{cet}

$^2A^- + ^2D^+ \xleftarrow{k_{fi}} {}^1SSRIP \underset{k_{ssrip}}{\overset{k_{crip}}{\rightleftharpoons}} {}^1CRIP$

k_{blet} $k_{cbet} + k_{em}$

$^1(A{+}D)$ $^1(AD)$

Scheme 4.

Table 2. Parameters for the Best Fit of the Kinetic Model Depicted in Scheme 3 Holding $k_{crip} = 0$

	$k_{cbet}+k_{em}$ ($\times 10^9$ s^{-1})	k_{ssrip} ($\times 10^9$ s^{-1})	k_{blet} ($\times 10^9$ s^{-1})	k_{fi} ($\times 10^9$ s^{-1})
ST/FN	7.67 ± 0.33	2.11 ± 0.15	0.439 ± 0.077	0.659 ± 0.110

In principle, the quenching of S_1 by FN may proceed by two pathways: when the molecules form a contact pair (k_{cet}) and when the molecules form a geminate pair (k_{let}). From the previous diffusional quenching studies of S_1 by FN,[29] the kinetics for the electron transfer under diffusion control are $2.63 \pm 0.23 \times 10^{10}$ s^{-1}. Employing the kinetic parameters in Table 2 to describe the dynamics of the CRIP and SSRIP in Scheme 4, the best fit to the experimental data occurs when $k_{cet} = 2.63 \pm 0.23 \times 10^{10}$ s^{-1} and $k_{let} = 0$. It was thus estimated that the diffusional quenching of S_1 by FN leads to greater than 90% CRIP formation.

These experiments reveal that for diffusional quenching of an excited singlet state by electron transfer where the driving force for the reaction is $\Delta G = -17.5$ kcal/mole in acetonitrile, the reaction occurs upon molecular contact producing CRIP. These findings are to be contrasted with the findings of Gould and Farid,[23] where for $\Delta G = -15$ kcal/mole, CRIP formation is not observed. Also, the present study brings into question the validity of deducing radical ion pair structures from the distance parameter contained in the time-dependent rate equations [Eq. (15)] used to analyze reaction dynamics in solution. The reaction distance for the diffusional quenching of S_1 by FN is 8.6 Å, which is consistent with the formation of a SSRIP. However, from the kinetic properties of the radical ion pair formed by the diffusional quenching of S_1 by FN, the predominate radical ion pair produced is the CRIP. Clearly great care must be exercised in deducing microscopic properties from theories that treat phenomena in terms of macroscopic parameters such as bulk diffusion coefficients.

4.2. Dynamics of Electron Transfer within the CRIP and SSRIP, k_{blet} and k_{cbet}

The free energy dependence of the rates of electron backtransfer within the SSRIP have been examined by Gould and Farid and their co-workers.[14,32] The SSRIP were formed by the fluorescence quenching of either 9,10-dicyanoanthracene (DCA) or 2,6,9,10-tetracyanoanthracene (TCA) by a series of alkyl-substituted benzenes in acetonitrile. The resulting SSRIP were assumed to decay by two pathways: electron backtransfer (k_{blet}) and separation to free radical ions (k_{fi}). To determine k_{blet}, the quantum yield of free radical ion formation was measured directly, and assuming a constant value for the rate of separation to free radical ions, $k_{fi} = 5 \times 10^8$ s^{-1}, the rate for electron backtransfer within a given SSRIP was deduced. As the ΔG for electron backtransfer was varied from –47.8 to –65.3 kcal/mole, the rate of electron backtransfer k_{blet}, decreased from 1.56×10^{10} s^{-1} to 7.76×10^8 s^{-1}. This kinetic behavior is characteristic of the Marcus "inverted region." Employing Eq. (6), the parameters V, λ_s, and λ_v were derived; these parameters were used in the calculation of the nonadiabatic curve shown in Figure 1.

Mataga and co-workers have examined the rate of electron backtransfer within the SSRIP, produced by diffusional quenching, where the ΔG was varied from –12.7 to –65.3 kcal/mole using a wide variety of donors and acceptors of varying electronic structure.[21] Over this range in ΔG, k_{blet} initially increased from 6.1×10^8 to 6×10^{10} s^{-1} and then decreased to 5×10^7 s^{-1}. This kinetic behavior is characteristic of the bell-shaped energy gap–dependence predicted by Eq. (6). However, caution must be exercised in attempting to derive the parameters V, λ_s, and λ_v from these kinetic data given the great variance in the electronic structure in the series of donors and acceptors. This is to be contrasted with the kinetic data of Gould and Farid, where the donor–acceptor pairs are of similar electronic structure.[14,32]

Charge recombination within the CRIP (k_{cbet}) has been studied by Asahi and Mataga.[20] The CRIP were produced by irradiation of the corresponding ground-state charge-transfer complex. The energy gap for the electron backtransfer decreased from –11.5 to –66.8 kcal/mole. As the energy gap decreased, the rate of electron backtransfer decreased from 5×10^{12} to 1×10^{10} s^{-1}. The bell-shaped dependence predicted by Eq. (7) was not observed, but rather the free energy dependence of k_{cbet} obeyed the following relationship:

$$k_{cbet} = \alpha \exp [-\beta \Delta G] \tag{19}$$

where α and β are constants. At present, there is no theoretical formulation that serves to rationalize this relationship.

4.3. Dynamics of Radical Ion Pair Diffusional Separation, k_{ssrip} and k_{fi}

Although there have been numerous studies examining the correlation between the dynamics of electron backtransfer within the CRIP and the SSRIP and the free energy change for the reactions, there have been only a limited number of studies that have had the sensitivity to examine the kinetics of CRIP separation from the SSRIP (k_{ssrip}) and SSRIP separation from FI (k_{fi}). Recently the question of how k_{ssrip} and k_{fi} depend on molecular structure has been addressed.

Mataga and co-workers have reported a femtosecond–picosecond study of the relaxation of CRIP in acetonitrile following irradiation of the charge-transfer complex between 1,2,4,5-tetracyanobenzene (TCB) and variety of aromatic hydrocarbons including benzene (Bz), toluene (Tl), *m*-xylene (*m*Xl), and *p*-xylene (*p*Xl).[18] Each of the radical ion pairs displays double-exponential kinetics, and the decays were modeled as depicted in Scheme 3. The results are given in Table 3.

Peters and co-workers have also examined the dependence of k_{ssrip} and k_{fi} for TS/FN, *trans*-4-chlorostilbene/fumaronitrile (TCS/FN), *trans*-stilbene/dimethylfumarate (TS/DF), and *trans*-4-chlorostilbene/dimethylfumarate (TCS/DF).[30] The results of these studies are given in Table 4.

Comparing the two kinetic studies, it is surprising to find that within series of structurally related compounds the rate of separation of CRIP to the SSRIP is independent of the molecular structure, while the rate of SSRIP separation to the FI does depend on the molecular structure. At present, there are no theoretical formulations for radical ion pair diffusion that will guide our understanding of the molecular structural dependence of these processes.

Table 3. Parameters for the Decay of CRIP and SSRIP for TCB/Aromatic Hydrocarbons in Acetonitrile

	k_{cbet}+k_{em} ($\times 10^9$ s^{-1})	k_{ssrip} ($\times 10^9$ s^{-1})	k_{blet} ($\times 10^9$ s^{-1})	k_{fi} ($\times 10^9$ s^{-1})
TCB/Bz	2.7[a]	4.5	0.68	1.50
TCB/Tl	5.7	4.3	1.10	1.60
TCB/*m*Xl	19.0	4.6	2.20	1.90
TCB/*p*Xl	22.0	4.6	5.10	2.60

Note: [a]No estimate of errors was given in this study.

Table 4. Parameters for the Decay of CRIP and SSRIP for Stilbene/Alkenes in Acetonitrile

	$k_{cbet}+k_{em}$ ($\times 10^9$ s^{-1})	k_{ssrip} ($\times 10^9$ s^{-1})	k_{blet} ($\times 10^9$ s^{-1})	k_{fi} ($\times 10^9$ s^{-1})
ST/DF	13.8 ± 1.42	2.01 ± 0.26	0.829 ± 0.222	0.859 ± 0.097
TCS/DF	9.29 ± 0.31	2.12 ± 0.16	0.469 ± 0.075	0.736 ± 0.127
ST/FN	7.67 ± 0.33	2.11 ± 0.15	0.439 ± 0.077	0.659 ± 0.110
TCS/FN	4.57 ± 0.24	2.00 ± 0.19	0.291 ± 0.087	0.516 ± 0.120

4.4. Dynamics of Intersystem Crossing within Radical Ion Pairs

The rate of singlet-to-triplet conversion for the SSRIP was first addressed by Weller.[5] Radical ion pairs produced on quenching of an excited singlet state by an electron donor are formed in a singlet state, ^{1}SSRIP. Through hyperfine interaction, the electron spins can be brought into a triplet alignment to form a triplet radical ion pair, ^{3}SSRIP. Electron backtransfer within this species may then produce an excited triplet state of the donor or acceptor molecule, providing it is energetically feasible.

Employing laser flash photolysis Weller measured the quantum yield of the formation of the pyrene triplet state following the quenching of the first excited singlet state of pyrene by dimethylanaline in acetonitrile.[5] Two mechanisms were identified for the formation of the pyrene triplet. A prompt pathway arises from intersystem crossing of the initially formed radical ion pair ^{1}SSRIP $\rightarrow$ ^{3}SSRIP with associated rate constant k_{st} followed by rapid electron backtransfer within the triplet radical ion pair to form triplet pyrene $^3A^*$, with rate constant k_{tet}. A delayed pathway for pyrene triplet formation results from the recombination of the free radical ions $^2A^- + {}^2D^+$, which on recombination form SSRIP, whose spin states are 75% triplet and 25% singlet. The resulting ^{3}SSRIP may undergo electron backtransfer to form triplet pyrene. The reaction scheme for the formation of prompt triplet pyrene is as in Scheme 5. The quantum yield for triplet pyrene formation from ^{1}SSRIP is given by

$$\Phi_T = k_{st}/(k_{fi} + k_{blet} + k_{st}) \quad (20)$$

which assumes that $k_{tet} >> k_{ts}$. Based on these experiments[5], the rate of intersystem crossing of the pyrene/dimethylanaline ^{1}SSRIP is $k_{st} = 6 \times 10^7$ s^{-1}. The rates of intersystem crossing in several other pyrene/amine (both aliphatic and aromatic) ^{1}SSRIP were also examined, and k_{st} was found to be independent of the nature of the amine.

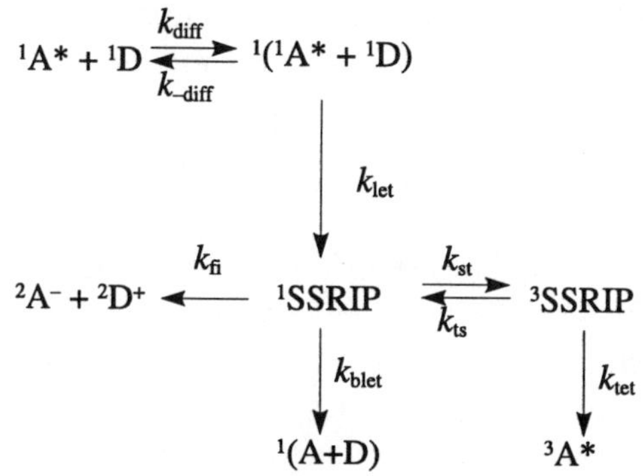

$$
\begin{array}{ccccc}
{}^1A^* + {}^1D & \underset{k_{-diff}}{\overset{k_{diff}}{\rightleftarrows}} & {}^1({}^1A^* + {}^1D) & & \\
 & & \downarrow k_{1et} & & \\
{}^2A^- + {}^2D^+ & \overset{k_{fi}}{\longleftarrow} & {}^1SSRIP & \underset{k_{ts}}{\overset{k_{st}}{\rightleftarrows}} & {}^3SSRIP \\
 & & \downarrow k_{blet} & & \downarrow k_{tet} \\
 & & {}^1(A+D) & & {}^3A^*
\end{array}
$$

Scheme 5.

There have been several studies revealing that triplet radical ion pairs participate in the ensuing chemistry subsequent to the irradiation of the charge transfer complex of *trans*-stilbene/fumaronitrile, TS/FN. Arnold and Wong[33] found that, upon irradiation of TS/FN in benzene, TS isomerizes to *cis*-stilbene (CS) and FN isomerizes to maleonitrile, where the *trans* to *cis* isomerization of stilbene is the more efficient reaction. These authors postulate that isomerization of both TS and FN occur through their respective triplet states, which are formed from the triplet radical ion pair produced on intersystem crossing of the singlet radical ion pair. This proposal received further support from the chemically induced nuclear spin polarization studies on TS/FN by Roth and Schilling.[34] Lewis and Simpson[35] undertook an extensive study of the effect of solvent and O_2 on the quantum yield for the isomerization of TS and FN and found that the predominant reaction is the isomerization of TS, as the product ratio of CS/maleonitrile is 15/1. In benzene the quantum yield for TS isomerization is approximately 0.10, which greatly diminished to 0.001 in acetonitrile. No bimolecular reaction products are observed in these studies.

Based upon the quantum yield studies of Lewis and Simpson,[35] Peters and Lee established[30] that the rate of intersystem crossing within the TS/FN singlet contact radical ion pair, $^1CRIP \rightarrow {}^3CRIP$, is $k_{st} = 1.7 \times 10^7$ s^{-1}. Given that rate of decay of the TS/FN ^1CRIP is approximately 10^{10} s^{-1}, the intersystem crossing to ^3CRIP represents only a minor pathway in the overall chemistry. However, in nonpolar solvents such as benzene, where the rate of the decay of ^1CRIP is approximately $10^8\ s^{-1}$,[26] chemistry within the triplet manifold makes a major contribution to the overall distribution of products. Thus the yield of products that arise from triplet radical ion pairs is very sensitive to the polarity of the solvent.

FUTURE DIRECTIONS

In recent years there has been significant progress in the development of our understanding of the parameters that control the dynamics and reaction pathways for radical ion pairs produced by the diffusional quenching of photoexcited molecules. With the advent of femtosecond–picosecond laser technology, the reaction dynamics of radical ion pairs can be measured directly. Development of theories of electron transfer has given us new insight into the molecular parameters that control the efficiency of electron transfer. However, there are at least three areas in the dynamics of radical ion pairs that need further investigation.

The first is further study, in terms of both theory and experiment, of the nature of the electron transfer process when the molecules come into molecular contact. At present it is not possible to predict the relative distribution between CRIP and SSRIP on diffusional quenching of an photoexcited molecule. This is most clearly exemplified in the Gould and Farid study of exciplex formation where only SSRIP are formed when $\Delta G = -15$ kcal/mole which is not consistent with the predictions derived from Figure 1. Either the molecular parameters, derived from charge recombination reactions, used to generate the adiabatic and nonadiabatic curves are grossly in error when applied to charge-separation reactions, or the theories resulting in Eqs. (6) and/or (7) are incomplete. Further kinetic studies for the formation of both CRIP and SSRIP are needed so that the parameters V, λ_s, and λ_v for charge separation can be obtained.

Another process that needs further examination is the rate of electron backtransfer within the CRIP, k_{cbet}. The adiabatic theory presented in Eq. (7) predicts a bell-shaped, rate-free energy relationship. However, the experiments of Mataga do not conform to this prediction. In order to further test the validity of the Jortner–Bixon theory of adiabatic electron transfer, more experimental studies of electron backtransfer within the CRIP are necessitated.

Finally, the effect of the solvent on the diffusional separation of radical ion pairs needs to be studied, again from the vantage point of experiment and theory. No theories at present provide a kinetic description of the effect of solvent or molecule structure on the diffusional separation process. Clearly the development of such theories would be greatly stimulated by a systematic experimental study of solvent effects on the dynamics of diffusional separation.

ACKNOWLEDGMENTS

This work is supported by a grant from the National Science Foundation, CHE 9120355. I would also like to thank the investigators who have worked on the various projects outlined in this review: Steve Angel, Erin O'Driscoll, Josh Goodman, John Simon, and Joseph Lee.

REFERENCES

1. Knibbe, H.; Rollig, K.; Schafer, F. P.; Weller, A. *J. Chem. Phys.* **1967**, *47*, 1184.
2. Knibbe, H.; Rehm, D.; Weller, A. *Ber. Bunsenges. Phys. Chem.* **1968**, *72*, 257.
3. Rehm, D.; Weller, A. *Isr. J. Chem.* **1970**, *8*, 259.
4. Marcus, R. A. *Ann. Rev. Phys. Chem.* **1964**, *15*, 155.
5. Weller, A. *Z. Phys. Chem. Neue Folge* **1982**, *130*, 129.
6. Weller, A. *Pure Appl. Chem.* **1982**, *54*, 1885.
7. Weller, A. *Z. Phys. Chem. Neue Folge* **1982**, *133*, 93
8. Kestner, N. R.; Jortner, J.; Logan, J. *J. Phys. Chem.* **1974**, *78*, 2148.
9. Fischer, S. F.; Van Duyne, R. P. *Chem. Phys.* **1977**, *26*, 9.
10. Jortner, J.; Bixon, M. *J. Chem. Phys.* **1988**, *88*, 167.
11. Barbara, P. F.; Walker, G. C.; Smith, T. P. *Science* **1992**, *256*, 975.
12. Newton, M.; Sutin, N. *Ann. Rev. Phys. Chem.* **1984**, *35*, 437.
13. Marcus, R. A. *Can. J. Chem.* **1959**, *37*, 155.
14. Gould, I. R.; Young, R. H.; Moody, R. E.; Farid, S. *J. Phys. Chem.* **1991**, *95*, 2068.
15. Yoshimori, A.; Kakitani, T.; Yoshitaka, E.; Mataga, N. *J. Phys. Chem.* **1989**, *93*, 8316.
16. Mataga, N.; Shioyama, H.; Kanda, Y. *J. Phys. Chem.* **1987**, *91*, 314.
17. Mataga, N.; Kanda, Y.; Asahi, T.; Miyasaka, H.; Okada, T.; Kakitani, T. *Chem. Phys.* **1988**, *127*, 239.
18. Ojima, S.; Miyasaka, H.; Mataga, N. *J. Phys. Chem.* **1990**, *94*, 7534.
19. Asahi, T.; Mataga, N.; Takahashi, Y.; Miyashi, T. *Chem. Phys. Lett.* **1990**, *171*, 309.
20. Asahi, T.; Mataga, N. *J. Phys. Chem.* **1989**, *93*, 6575.
21. Mataga, N.; Asahi, T.; Kanda, Y.; Okada, T.; Kakitani, T. *Chem. Phys.* **1988**, *127*, 249.
22. Kikuchi, K.; Niwa, T.; Takahashi, Y. Ikeda, H.; Miyashi, T.; Hoshi, M. *Chem. Phys. Lett.* **1990**, *173*, 421.
23. Gould, I. R.; Mueller, L. J.; Farid, S. *Z. Phys. Chem.* **1991**, *170*, 143.
24. Gould, I. R.; Farid, S. *J. Phys. Chem.* **1992**, *96*, 7635.
25. Goodman, J. L.; Peters, K. S. *J. Am. Chem. Soc.* **1985**, *107*, 314.
26. O'Driscoll, E.; Simon, J. D.; Peters, K. S. *J. Am. Chem. Soc.* **1990**, *112*, 7091.
27. Angel, S.; Peters, K. S. *J. Phys. Chem.* **1989**, *93*, 713.
28. Angel, S. A.; Peters, K. S. *J. Phys. Chem.* **1991**, *95*, 3606.

29. Peters, K. S.; Lee, J. *J. Phys. Chem.* **1992**, *96*, 8941.
30. Peters, K. S.; Lee, J. *J. Am. Chem. Soc.* **1993**, *115*, 3643.
31. Collins, F. C.; Kimball, G. *J. Colloid Sci.* **1949**, *4*, 425.
32. Gould, I. R.; Ege, D.; Moser, J. E.; Farid, S. *J. Am. Chem. Soc.* **1990**, *112*, 4290.
33. Arnold, D. R.; Wong, P. C. *J. Am. Chem. Soc.* **1979**, *101*, 1894.
34. Roth, H. D.; Schilling, M. L. M. *J. Am. Chem. Soc.* **1980**, *102*, 4303.
35. Lewis, F. D.; Simpson, J. T. *J. Phys. Chem.* **1979**, *83*, 2015.

DISSOCIATIVE ELECTRON TRANSFER

Jean-Michel Savéant

Advances in Electron Transfer Chemistry
Volume 4, pages 53–116.

ISBN: 1-55938-506-5

1. INTRODUCTION

Even when it involves outer-sphere electron donors, electron transfer to organic or inorganic molecules is often accompanied by the breaking of a chemical bond. By an "outer-sphere" electron donor, we mean an electron donor that is able to transfer one electron to another molecule without breaking an existing bond or forming a new bond within its own molecular structure. Under these conditions, one important question is whether the electron transfer step and the bond-breaking step occurring in the acceptor molecule are successive or concerted. The kinetics of the overall reaction are different in each case.

In the first case, the electron transfer process is of the outer-sphere type, and two limiting kinetic situations may be reached according to the outcome of the competition between the reverse electron transfer and the bond-breaking step. If the latter is slower than the former, the rate-determining step is the bond-breaking step and the electron transfer interferes with the kinetics of the overall reaction solely through its equilibrium thermodynamic properties. In the opposite case, the forward electron transfer step is rate determining.

If electron transfer and bond breaking are concerted, the kinetics is that of a single step that possesses an inner-sphere character and whose dynamics is governed not only by solvent reorganization and internal changes in lengths and angles of the bonds that are not broken but also by a contribution from the bond breaking itself.

The fact that the reaction dynamics are different in the sequential and concerted mechanisms has important implications in electron transfer chemistry. In the common case, where the acceptor is a neutral closed-shell species, a radical and an anionic fragment are formed on reductive cleavage. The possibility of triggering a radical chemistry by transferring one electron from an outer-sphere source thus depends on the competition between the target reaction and side-reactions that destroy the radical. Among the latter, conversion of the radical into the corresponding anion by electron transfer from the same electron donor used to produce the radical is of particular importance. Thus, both the dynamics of the electron transfer producing the radical and that of its electron transfer reduction are important in defining strategies that may be used to trigger, by means of outer-sphere electron donors, either a radical chemistry or an ionic chemistry.

Radicals generated by electron transfer from an outer-sphere donor, whether homogeneous or heterogeneous (i.e., electrochemically at an inert electrode), are often, though not always, better electron acceptors than the substrate from which they are produced. Under such conditions, if the reductive cleavage follows a concerted mechanism there is no hope of triggering a radical chemistry electrochemically since the radical is produced at the electrode surface at a potential such that it is immediately reduced into the corresponding anion. If, oppositely, a stepwise mechanism is followed, the radical is formed farther from the electrode surface and may thus react with an appropriate reagent before coming back to that surface and being reduced there, according to what electrochemists call an "ECE" (E : electron transfer, C : chemical reaction) mechanism.[1] In the homogeneous case, the reduction of the radical is very fast, practically at the diffusion limit, and competes with two different types of generation kinetics according to the character of the reductive cleavage mechanism.

The $S_{RN}1$ substitution reaction at sp^2 or sp^3 carbons[2]

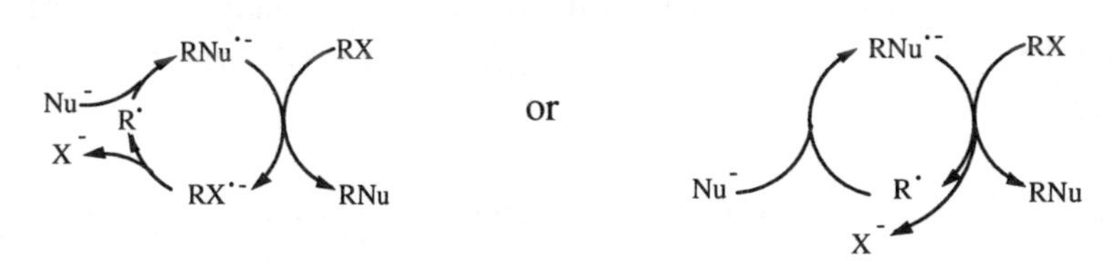

is a typical example of a radical chemistry where the abovementioned problems are of key importance for the target substitution to overcome the competing hydrogenolysis reactions.

In the case where the reductive cleavage follows the sequential mechanism and is governed by the initial (outer-sphere) electron transfer step, the dynamics of the reaction may be described approximately, but generally, by the Marcus–Hush model.[3] The activation-driving force relationship is quadratic,

$$\Delta G^{\neq} = w_R + \Delta G_0^{\neq}\left(1 + \frac{\Delta G^0 - w_R + w_P}{4\Delta G_0^{\neq}}\right)^2 \tag{1}$$

for the forward reaction and

$$\Delta G^{\neq} = w_P + \Delta G_0^{\neq}\left(1 - \frac{\Delta G^0 - w_R + w_P}{4\Delta G_0^{\neq}}\right)^2 \tag{2}$$

for the reverse reaction. ΔG^0: standard free energy of the reaction (i.e., the opposite of the driving force in terms of free energy); $\Delta G_0^{\neq}$: standard activation free energy (i.e., the activation free energy of the forward and reverse reactions at zero driving force, or in other words, the intrinsic barrier free energy); w_R: work required for bringing the reactants from infinity to reacting distance (i.e., to form the "precursor complex"); w_P: work required to form the "successor complex", from infinity to the reacting distance. The intrinsic barrier is the sum of an external (solvent) and internal (bond lengths and angles) reorganization factors λ_0 and λ_i:

$$\Delta G_0^{\neq} = \frac{\lambda_0 + \lambda_i}{4} \tag{3}$$

which can be estimated from the dielectric properties of the solvent and from the force constants and length or angle variations of the bonds that undergo a significant change upon electron transfer, respectively. The quadratic character of the activation–driving force relationship implies that the transfer coefficient (symmetry factor) α varies with the driving force:

$$\alpha = \frac{\partial \Delta G^{\neq}}{\partial \Delta G^0} = \frac{1}{2}\left(1 - \frac{\Delta G^0 - w_R + w_P}{4\Delta G_0^{\neq}}\right) \tag{4}$$

(assuming that the work terms do not vary significantly with the driving force). The above equations apply for homogeneous as well as heterogeneous (electrochemical) outer-sphere electron transfer reactions.

In the case of dissociative electron transfer (i.e., the case in which electron transfer and bond breaking are concerted) several assumptions underlying the preceding model of outer-sphere electron transfer cease to be applicable. Two questions then arise: What is the form of the activation–driving force relationship [may Eq. (1) still be employed?] and What is the contribution of bond breaking to the activation energy? Section 2 is devoted to answering these questions.

Recognition that a reductive cleavage reaction follows a stepwise mechanism is not too difficult a task when the intermediate resulting from the uptake of one electron has a lifetime long enough to fall within the time window of available kinetic techniques. The problem of distinguishing between stepwise and concerted mechanisms is more arduous when the intermediate is so short-lived that kinetic control occurs by electron transfer in both cases. These questions will be addressed in

Section 3, in which we will also review and analyze the main factors that caused the mechanism to be of one type or the other.

A large number of chemical reactions are classically viewed as involving the transfer of an electron pair rather than of a single electron. A typical example of such problems is the S_N2 nucleophilic substitution. However the question arises of whether an S_N2 reaction

$$\mathrm{Nu:^- + RX \rightarrow RNu + X:^-}$$

might consist of the succession of one-electron transfer, bond-breaking, and bond-forming steps:

$$\mathrm{Nu:^- + RX \rightleftharpoons RX^{\bullet -} + Nu^{\bullet}}$$

$$\mathrm{RX^{\bullet -} \rightleftharpoons R^{\bullet} + X:^-}$$

$$\mathrm{R^{\bullet} + Nu^{\bullet} \rightarrow RNu}$$

$$\overline{\mathrm{Nu:^- + RX \rightarrow RNu + X:^-}}$$

In analyzing such questions, it is obviously important to know whether the two first steps are concerted or successive. More generally, even if the three steps above are concerted (i.e., the reactions follows a classical S_N2 mechanism), it is interesting to know if the reaction dynamics is better represented by a mechanism that involves concertation of a single electron transfer with bond-breaking and bond formation or rather the transfer of an electron pair (i.e., the displacement of a carbonium ion R^+ from X^- to Nu^-. These questions that have already been discussed (see Ref. 2s and Refs. therein) will be addressed here in terms of the transition between outer-sphere dissociative electron transfer and S_N2 substitution in the case of the reaction of aliphatic halides with aromatic anion radicals.

2. DYNAMICS OF DISSOCIATIVE ELECTRON TRANSFER

2.1. A General and Approximate Model

In the modelling of a reaction of the type

$$\mathrm{RX} + e^- \rightleftharpoons \mathrm{R^{\bullet} + X^-} \text{ (electrochemical)} \tag{1}$$

$$\mathrm{RX + D^{\bullet -} \rightleftharpoons R^{\bullet} + X^- + D} \text{ (homogeneous)} \tag{1'}$$

in a polar medium, where $D^{\bullet -}$ is an outer-sphere homogeneous electron donor and RX does not necessarily represent an alkyl halide but, more generally, any electron acceptor undergoing dissociative electron transfer, the harmonic oscillator approximation used in Marcus–Hush theory of outer-sphere electron transfer to describe the internal reorganization of reactants and products is not suited for describing the broken bond in the product system. The following Morse curve–based model has been proposed to account for the contribution of bond breaking to the activation barrier while also taking account of solvent reorganization and changes in lengths or angles of the bonds not being broken during electron transfer.[4a]

As in Marcus–Hush theory, the Born–Oppenheimer approximation is assumed to hold and the reaction to be adiabatic. The potential energy surfaces for the reactants and products depend on three types of reaction coordinates: the stretching of the R–X bond from equilibrium; z, the same fictitious charge x (which varies from x_R to $x_p = x_R + 1$ during the reaction), describing the solvent fluctuational configurations as in Marcus theory of outer-sphere electron transfer; and the vibration coordinates y (which vary from y_R to y_P during the reaction) of the bonds that are not broken during the reaction. The corresponding contributions to the free energy of the reactant and product systems are regarded as independent and additive. The two last contributions are treated in exactly the same way as in Marcus–Hush theory. As regards bond breaking, the potential energy of the reactants is assumed to depend on the R–X distance according to the RX Morse curve, and the potential energy curve for the products is assumed to be purely dissociative and to be the same as the repulsive part of the reactant Morse curve (see Figure 1). The assumption that the repulsive portions of the reactants' and products' potential energy curves are the same has previously been made in the interpretation of the kinetics of gas-phase thermal electron attachment to alkyl halides.[4b] It was based on the approximation that the repulsive term arises from interactions involving the core electrons and the nuclei and should therefore not be dramatically affected by the presence of an additional peripheral electron. Possible attractive interactions in the product system, such as induced dipole-charge and quadrupole-charge interactions between $R^{\bullet}$ and X^{-}, are regarded as small, smaller than in the gas phase, owing to the presence of a surrounding polar medium. Under these conditions, the free energies may be expressed as

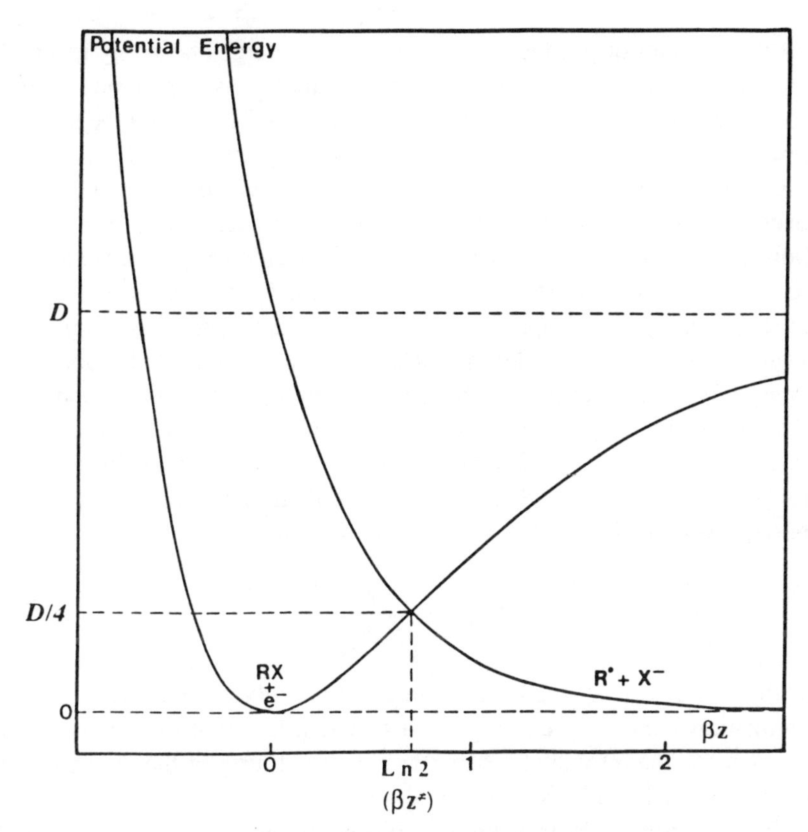

Figure 1. Morse curves of the reactants and products at zero driving force (z: elongation of the RX distance from equilibrium, $\beta = \nu_0(2\pi^2\mu/D)^{1/2}$, with ν_0 representing the vibration frequency, μ the reduced mass of the two atoms of the R-X bond, and D the RX bond dissociation energy.

$$G_R = G_R^0 + w_R + \tfrac{1}{2}\lambda_0(x - x_R)^2 + \tfrac{1}{2}\sum f_R(y - y_R)^2$$
$$+ D\{1 - \exp[-\beta(z - z_R)]\}^2 \qquad (5)$$

for the reactants and as

$$G_P = G_P^0 + w_P + \tfrac{1}{2}\lambda_0(x - x_P)^2 + \tfrac{1}{2}\sum f_P(y - y_P)^2$$
$$+ D\{\exp[-\beta(z - z_R)]\}^2 \qquad (6)$$

for the products. G_R^0 and G_P^0 are the equilibrium free energies of the reactant and product systems, respectively, x is a fictitious charge borne by the reactant or product, x_R and x_P are the charges borne by reactant and product at their equilibrium states, and λ_0 is the same solvent reorganization factor as in Marcus–Hush theory. The y's are the lengths or angles of the bonds not being broken during the reaction, y_R and y_P their equilibrium values; f_R and f_P are the corresponding harmonic oscillator force constants (the summations are extended over all such bonds). z is the length of the bond being broken during the reaction, z_R its equilibrium value in the reactant system, D the bond dissociation energy, and β the shape factor defined in the caption of Figure 1. Determination of the saddle point on the intersection of the potential energy hypersurfaces defined by Eqs. (5) and (6) leads to the conclusion that the activation–driving force relationships have exactly the same quadratic form as in Marcus–Hush theory of outer-sphere electron transfer [Eqs. (1) and (2)] and that the standard activation free energy (intrinsic barrier) is given by

$$\Delta G_0^{\neq} = \frac{D}{4} + \frac{\lambda_0}{4} + \frac{\lambda_i}{4} \tag{7}$$

with λ_0 and λ_i being the usual Marcus–Hush solvent and internal reorganization factors. According to this model, the contribution of bond breaking to the intrinsic barrier is simply one fourth of the dissociation energy of the bond being broken.

The location of the transition state is given by

$$(z^{\neq} - z_R) = \frac{1}{\beta}\left[\ln 2 - \ln(1 - \frac{\Delta G^0}{4\Delta G_0^{\neq}})\right] \tag{8}$$

The quadratic character of the activation–driving force relationships implies here, too, that the transfer coefficient (symmetry factor) α varies with the driving force, being smaller than 0.5 at large driving forces and larger than 0.5 at small driving forces.

Recent quantum-chemical *ab initio* calculations in the methyl halide series have confirmed the validity of the Morse curve–based empirical model in terms of predicted activation free energy and geometry of the transition state (see Table 1).[4c]

So far, in the application of the model to the analysis of heterogeneous and homogeneous reactions, the pre-exponential factor relating the rate constant to the activation free energy has been taken as equal to the heterogeneous or bimolecular collision frequencies, as in the original version of Marcus theory[3a–c]

Table 1. Transition-State Characteristics of the Dissociative Electron Transfer
$CH_3X + e^- \rightarrow CH_3^{\bullet} + X^-$
Comparison of the Predictions of the Empirical Model with Quantum-Chemical Calculations

	$\Delta G^{\neq\, a}$			$z^{\neq} - z_R$ [b]		
	Ab initio	*Empirical model*		*Ab initio*	*Empirical model*	
X	*calculations*	*c*	*d*	*calculations*	*c*	*d*
F	49.1	51.9	43.2	0.59	0.67	0.56
Cl	23.5	22.7	19.3	0.42	0.48	0.42
Br	15.9	17.5	12.5	0.37	0.45	0.36
I	9.5	14.6	7.8	0.33	0.41	0.30

Notes: [a]Activation free energy in Kcal / mol.
[b]Increase of the C-X distance from the ground state, in Å. In the empirical model, $z^{\neq}$ is calculated from Eq. (8).
[c]D and ΔG^0 from the *ab initio* calculations.
[d]From the experimental D and ΔG^0 values.

$$k = Z \exp\left(-\frac{\Delta G^{\neq}}{RT}\right) \tag{9}$$

with

$$Z^{\text{el}} = \left(\frac{RT}{2\pi M}\right)^{\frac{1}{2}} \tag{10}$$

(M : molar mass), in the electrochemical case, and

$$Z^{\text{hom}} = (a_1 + a_2)^2 \left(\frac{8\pi RT}{\mu}\right)^{\frac{1}{2}} \tag{11}$$

(μ : reduced molar mass of the two reactants; a_1, a_2: hard-sphere equivalent radii of the two reactants, in the homogeneous case).

In fact, the identification of Z with the collision frequency is certainly a crude approximation, and the same effects that have been discussed in the case of outer-sphere electron transfer[5] should be taken into account here.[4d] A better expression for the rate constant is thus[5a,b]

$$k = K_A\, \kappa_e\, \nu_n \exp\left(-\frac{\Delta G^{\neq}}{RT}\right) \tag{12}$$

where K_A is the equilibrium constant for the formation of the precursor complex, ν_n is the effective nuclear frequency, and κ_e is the electronic transmission factor that includes the possible nonadiabaticity effects as a function of the electronic coupling matrix element. The latter energy (i.e., the resonance or avoided crossing energy) should also be subtracted from the values of $\Delta G^{\neq}$ derived from the diabatic potential free energy surfaces in the procedure described above.

The effective nuclear frequency ν_n is expected to be dominated by the vibration frequency of the cleaving bond, which is generally larger than that of the solvent mode. Under such conditions, the dynamic of solvent fluctuations is expected to govern the effective value of the pre-exponential factor:

$$k = K_A\, \kappa_e\, \nu \exp\left(-\frac{\Delta G^{\neq}}{RT}\right) \tag{13}$$

In the framework of a steady-state approximation, ν can be expressed as[5d]

$$\frac{1}{\nu} = \frac{1}{\nu_n} + \frac{F\left(\dfrac{D+\lambda_i}{D+\lambda_i+2\lambda_0}, \Delta\dfrac{G^{\neq}}{RT}\right)}{\tau_L} \tag{14}$$

(τ_L: solvent longitudinal relaxation time). From available numerical computation of the function F, it is seen that there is an approximate mutual compensation for the effect of $(D + \lambda_i)/\lambda_0$ and of $\Delta G^{\neq}/RT$ on the values of F.

It follows that a viable strategy for minimizing the impact of these various effects in the aim of focussing attention on the contribution of bond breaking to the activation barrier consists of comparing the kinetic data obtained with the investigated dissociative electron transfer reaction to those pertaining to an outer-sphere reaction with little internal reorganization taking place in the same medium, rather than attempting an absolute comparison between the model and the experimental rate data. Such outer-sphere reactions involve small values of both λ_i/λ_0 and $(\Delta G^{\neq}/RT)$, whereas dissociative electron transfers are characterized by large values of $(D + \lambda_i)/\lambda_0$ and of $(\Delta G^{\neq}/RT)$. This comparative strategy also possesses the advantage of allowing a more reliable estimate of the solvent reorganization factor λ_0 than does the strict application of the Marcus–Hush equations based on the Born model of solvation. Thus, taking into account the variation of K_A from one reaction to the other, which is tantamount to that of Z, and the variation of λ_0 related to that of the hard-sphere equivalent radius, the old Z-based formalism may still be used under the assumption that the resonance energy and the other factors that govern the pre-exponential factor are about the same in the two reactions. Examples of the application of this strategy are described in Sections 2.2 and 2.4.

2.2. Electrochemical Testing of the Model

As will be made clear in Section 3, the reductive cleavage of simple alkyl halides by outer-sphere heterogeneous or homogeneous electron donors in polar solvents follows the concerted rather than the stepwise mechanism. It thus offers an example where the validity of the above model of dissociative electron transfer could be tested experimentally. This test has indeed been performed in the series of butyl (*n*-, *sec*-, *tert*-) halides (iodides, bromides, chlorides) in *N,N'*-dimethylformamide (DMF) as a solvent using both heterogeneous and homogeneous electron donors.[4a,d;6] For the electrochemical reduction of alkyl halides, the best approximation of a heterogeneous outer-sphere electron donor was found to be a glassy carbon electrode.

The comparison between the predictions of the dissociative electron transfer theory and the experimental data is summarized in Table 2. More precisely, the value of the intrinsic barrier free energy derived from the experimental data (namely, the cyclic voltammetric peak potential, E_p, or the midpoint between the peak and the half-peak potential, E_m), $\Delta G_0^{\neq}$ (exp^{al}) is compared to the value obtained from Eq. (7) (in which λ_i is neglected), $\Delta G_0^{\neq}$ (theor). The neglect of λ_i amounts to assuming that the energy required to change the bond angles in the reacting carbon, going from a tetrahedral to a planar configuration, is included in the Morse-curve description. To obtain $\Delta G_0^{\neq}$ (exp^{al}), we apply Eq. (1) in a form suited to electrochemical reactions:

$$\Delta G^{\neq} = \Delta G_0^{\neq}\left[1 + \frac{E_{p,m} - (E^0 + \phi_r)}{4\Delta G_0^{\neq}}\right]^2 \qquad (15)$$

for the forward reaction, in which $E_{p,m}$ designates E_p or E_m according to the case; E^0, the standard potential of the $RX + e^- \rightleftharpoons R^{\bullet} + X^-$ couple; and ϕ_r the potential difference between the reacting site and the solution. The reacting site is generally assumed to be located in the outer Helmoltz plane of the electrochemical double layer,[7a] and ϕ_r is obtained from previous measurements carried out in the same solvent (here DMF) with the same supporting electrolyte cation (here $n\text{-}Bu_4^+$).[7b] In the potential region of interest, the variation of ϕ_r with the electrode potential E is with good precision linear. In volts,

$$\phi_r = 0.011\,E - 0.052 \tag{16}$$

$\Delta G^{\neq}$, the activation free energy of the forward reaction, has two slightly different values, $\Delta G^{\neq}_m$ and $\Delta G^{\neq}_p$, according to whether it is derived from E_m or E_p:

$$\Delta G^{\neq}_m = \frac{RT}{F}\ln\left[Z^{\mathrm{el}}\left(\frac{RT}{\alpha F v D}\right)^{\frac{1}{2}}\right] + 0.145\frac{RT}{F} \tag{17}$$

$$\Delta G^{\neq}_p = \frac{RT}{F}\ln\left[Z^{\mathrm{el}}\left(\frac{RT}{\alpha F v D}\right)^{\frac{1}{2}}\right] - 0.780\frac{RT}{F} \tag{18}$$

where D is the diffusion coefficient of RX, v the scan rate, and α the transfer coefficient. In the establishment of Eqs. (17) and (18), the quadratic character of the activation–driving force relationship and thus the variation of the transfer coefficient with the electrode potential was neglected.[4a,8] Such a simplification is perfectly legitimate for each scan rate individually since the potential difference between the foot and the peak of each individual voltammogram is small. In the derivation of $\Delta G^{\neq}_{m,p}$, α is obtained from the experimental data or, better, from predicted values as defined later on and $\Delta G^{\neq}_{m,p}$ obtained along a rapidly converging iterative procedure. Once $\Delta G^{\neq}_{m,p}$ is known, $\Delta G^{\neq}_0$ (exp^{al}) is obtained from Eq. (12) as

$$\Delta G^{\neq}_0(\mathrm{exp}^{\mathrm{al}}) = \frac{\left[(E_{m,p} - E^0 - \phi_r - 2\Delta G^{\neq}_{m,p})^2 - (E_{m,p} - E^0 - \phi_r)^2\right]^{\frac{1}{2}}}{4} - \frac{(E_{m,p} - E^0 - \phi_r - 2\Delta G^{\neq}_{m,p})}{4} \tag{19}$$

In this estimate, E^0 is calculated from literature thermochemical data with attention to the particular electrode reference used (here the aqueous saturated calomel electrode):

Table 2. Cyclic Voltammetric Reductive Cleavage of Butyl Bromides and Iodides[a]; Comparison between the Predictions of the Theory of Dissociative Electron Transfer and the Experimental Data

	Butyl iodides						Butyl bromides					
Quantity	*n-BuI*		*s-BuI*		*t-BuI*		*n-BuBr*		*s-BuBr*		*t-BuBr*	
$-E_m$ [b]	2.252		1.982		1.839		2.756		2.540		2.397	
$-\phi_r$ [b]	0.090		0.084		0.081		0.102		0.097		0.090	
Z^{el}(cm/s)			4.6×10^3						5.2×10^3			
$\Delta G^{\neq}$ [c,d]	0.360		0.346		0.347		0.366		0.354		0.356	
$\Delta G_{RX/R^{\bullet}}$ [d]	0	0.112	0	0.112	0	0.112	0	0.121	0	0.121	0	0.121
$-E^0$ [b]	1.075	1.209	0.946	1.080	0.826	0.960	1.109	1.230	1.049	1.190	0.929	1.050
$\Delta G_0^{\neq}$ [d] (exp[al])	0.813	0.762	0.754	0.703	0.740	0.688	0.951	0.908	0.919	0.867	0.915	0.871
a [e]	3.56		3.58		3.62		3.50		3.50		3.54	
a [e]	3.04		3.05		3.06		2.82		2.82		2.83	
λ_0 [d]	0.685		0.682		0.679		0.735		0.735		0.731	
D [d,f]	2.56	2.67	2.38	2.49	2.20	2.31	3.00	3.12	2.95	3.07	2.87	2.99
$\Delta G_0^{\neq}$ [d] (theor)	0.811	0.845	0.766	0.800	0.719	0.753	0.934	0.964	0.921	0.951	0.906	0.936
α(pred)	0.33	0.34	0.33	0.34	0.34	0.36	0.29	0.30	0.31	0.32	0.31	0.32
α(theor)	0.33	0.36	0.33	0.36	0.34	0.37	0.29	0.32	0.31	0.34	0.31	0.33
α(exp[al])	0.30		0.33		0.32		0.25		0.25		0.20	

Notes: [a]In DMF + 0.1 M n-Bu_4NBF_4 at a glassy carbon electrode at 25 °C; scan rate: 0.1 V/s.
[b]In V versus SCE.
[c]$D = 10^{-5}$ cm^2 s^{-1}.
[d]In eV.
[e]In Å.
[f]From literature thermochemical data (see text).

$$E^0 = \mu^*_{RX} - \mu^*_{R^\cdot} - \mu^*_{X^-}$$

(the μ^*s are the standard chemical potentials of the subscript compounds). While $\mu^*_{X^-}$ is available for solvents currently used in electrochemistry (here DMF), μ^*_{RX} and $\mu^*_{R^\cdot}$ are available only in the gas phase. In the passage from the gas phase to the solvent, one may assume that the variations of μ^*_{RX} and $\mu^*_{R^\cdot}$ are about the same (the solvation free energy of the permanent and induced R – X dipole would be nearly the same as that of the induced $R^\bullet$ dipole). A first estimate of E^0 is thus obtained from the μ^*_{RX} and $\mu^*_{R^\cdot}$ gas-phase literature values and from the solvation free energy of X^-.[8] Another approach involves the following approximations:[6,9] The variation of μ^*_{RX} and $\mu^*_{R^\cdot}$ would be the same as that of CH_3X and CH_4, respectively, independently of the nature of the polar solvent considered. A second estimate of E^0 is thus obtained from the same thermochemical data[10a] and the known solvation free energies of CH_3X and CH_4 in water.[9] The second values are thus less than the first by 0.134, 0.121, and 0.112 V for I, Br, and Cl respectively.[6]

To assign a value to $\Delta G_0^{\neq}$(theor) we need an estimate of the solvent reorganization factor λ_0^{het} for the heterogeneous reaction. In the theory of outer-sphere electron transfer, there are two estimates of λ_0^{het}, both being based on a hard-sphere approximation of the molecules and a Born-type description of solvation. In the Marcus version,[3a] the reactant is located at a distance to the electrode equal to its equivalent hard-sphere radius a, the double-layer correction is explicitly introduced as an estimate of the work terms, and image force effects are taken into account. Then

$$\lambda_0 - \left(\frac{1}{D_{op}} - \frac{1}{D_S}\right)\frac{1}{4a} \tag{20}$$

(D_{op}, D_S : dynamic and static dielectric constant of the solvent, respectively). In Hush's version, the reaction site is considered to be far enough from the electrode surface for the double-layer correction and the image force effect to be neglected. Thus

$$\lambda_0 = \left(\frac{1}{D_{op}} - \frac{1}{D_S}\right)\frac{1}{2a} \tag{21}$$

The truth probably lies in between these two extreme approximations. The best way to handle this problem is to examine the experimental data available. The experiments carried out with a series of aromatic molecules giving rise to stable anion radicals (at least within the time window of the experimental techniques) indicate a roughly linear variation of

$\Delta G_0^{\neq}$ (exp[al]) with $1/a$,[6b] a being determined from the density ρ of the compound according to

$$a(\text{Å}) = 10^8 \left(\frac{3M}{4\pi N_A \rho} \right)^{\frac{1}{3}} \tag{22}$$

It is thus found that

$$\Delta G_0^{\neq}(\text{eV}) = \frac{2.08}{a\,(\text{Å})} \tag{23}$$

The reduction of these compounds offers the best approximation of an outer-sphere electron transfer reaction where the internal reorganization factors can be neglected in view of the robustness of the aromatic framework and thus where $\Delta G_0^{\neq} = \lambda_0^{\text{het}}/4$. The empirical equation

$$\lambda_0(\text{eV}) = \frac{8.31}{a\,(\text{Å})} \tag{24}$$

may thus be used to estimate the solvent reorganization factor for theoretical predictions concerning double-layer corrected rate data. The value thus obtained may not represent the actual solvent reorganization factor for this series of outer-sphere electron transfer reaction. It may also contain a factor representing the ratio between the collision frequency Z^{el} and the actual pre-exponential factor. As discussed in Section 2.1, this is of little importance since, in the framework of the comparative strategy, the combination of the two factors is precisely what is needed in the analysis of the dissociative electron transfer data.

In the present case there is an additional problem for the estimation of λ_0^{het}; namely, the fact that in RX molecules, the charge variation that triggers the solvent reorganization essentially involves the X portion of the molecule, this being partly hindered by the presence of the R residue. In this connection, the RX molecule may be approximated by two tangent spheres as represented in Figure 2, a_X being taken as the ionic radius of X^-.

In previous discussions of this problem,[2s,4a] the radius a to be used in Eq. (24) was taken as the arithmetic mean of a_{RX} and a_X. A probably better approximation of the charging of this ensemble of spheres consists in taking for a

$$a = \frac{a_X(2a_{RX} - a_X)}{a_{RX}} \tag{25}$$

as done in the results reported in Table 2.

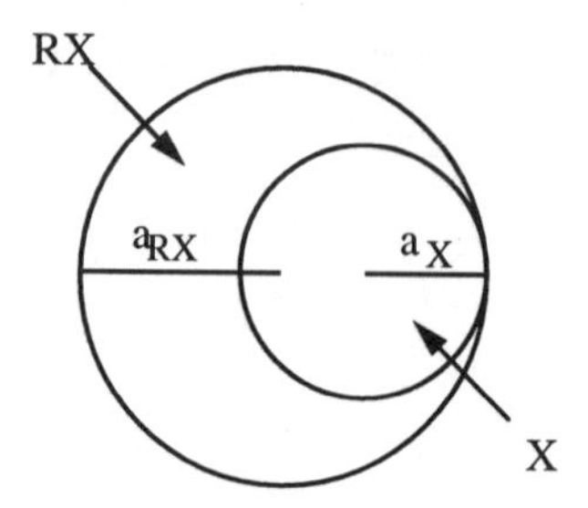

Figure 2. Approximation of the RX molecule for estimating the solvent reorganization factor.

When estimating the value of the bond dissociation energy to be used in applying Eq. (7), the gas-phase value of the formation enthalpy of $X^{\bullet}$ is used in all cases. In the case where E^0 is estimated from the gas-phase values of μ^*_{RX} and $\mu^*_{R^{\bullet}}$, D is accordingly calculated from the gas-phase formation enthalpies of RX and $R^{\bullet}$. In the case where the difference in the solvation free energies of RX and $R^{\bullet}$ is approximated by the difference between CH_3X and CH_4, the same amount is added to the gas-phase value of D.

The results displayed in Table 2 show that the theoretically predicted intrinsic barriers are in quite good agreement with the experimental data when E^0 and D are estimated on the basis of RX and $R^{\bullet}$ gas-phase data. The agreement is not as good with the second estimate of E^0 and RX.

Using the same experimental data, one can test another aspect of the dissociative electron transfer theory, namely the quadratic character of the activation–driving force relationship on comparing the experimental values of the transfer coefficient α (exp^{al}) with the predicted α (pred) and theoretical α (theor) values. α (pred) is the value obtained from Eq. (26), which is the adaptation of Eq. (4) to electrochemical reactions:

$$\alpha = \frac{1}{2}\left(1 + \frac{E_{p,m} - E^0 - \phi_r}{4\Delta G_0^{\neq}}\right) \tag{26}$$

taking for $\Delta G_0^{\neq}$ the experimental value $\Delta G_0^{\neq}$ (exp^{al}) obtained from the experimental E_m data by means of Eq. (19), whereas α (theor) is obtained from Eq. (26) using for $\Delta G_0^{\neq}$ the theoretical value derived from Eq. (7) as explained above. It is seen from Table 2 that, although the experimental values of α are somewhat smaller than predicted, the striking fact that they lie well below 0.5 is nicely explained by the quadratic character of the activation–driving force relationship, taking into account that the

Table 3. Comparison between the Predictions of the Theory of Dissociative Electron Transfer and the Experimental Data for the Cyclic Voltammetric Reductive Cleavage of Benzyl Chloride and Bromide[a]

Compound		*PhCH₂Cl*		*PhCH₂Br*	
$E_p^{\ b}$	0.1 V/s	–2.21		–1.71	
	1.0 V/s	–2.30		–1.82	
	10.0 V/s	–2.41		–1.94	
$\phi_r^{\ b}$	0.1 V/s	–0.089		–0.078	
	1.0 V/s	–0.091		–0.080	
	10.0 V/s	–0.094		–0.082	
Z^{el}	(cm/s)	5.6×10^3		4.8×10^3	
$\Delta G^{\neq\ c,d}$	0.1 V/s	0.347		0.342	
	1.0 V/s	0.318		0.313	
	10.0 V/s	0.289		0.285	
$\Delta G^0_{RX/R^\bullet}\ ^d$		0	0.112	0	0.121
$E^{0\ b}$		–0.760	–0.872	–0.635	–0.756
$\Delta G_0^{\neq}$ (exp[al])[d]	0.1 V/s	0.899	0.858	0.759	0.713
	1.0 V/s	0.897	0.856	0.766	0.721
	10.0 V/s	0.899	0.860	0.776	0.731
	Av.	0.898	0.858	0.767	0.722
a_X, a_{RX},	$\lambda_0\ ^d$	1.81, 3.58,	1.96, 3.61,	0.729	
a^e		2.70, 0.767	2.86,		
$D^{d,f}$		2.99	3.10	2.30	2.42
$\Delta G_0^{\neq}$ (theor) $= \frac{D+\lambda_0}{4}\ ^d$		0.939	0.967	0.757	0.788
α (pred)	0.1 V/s	0.311	0.318	0.336	0.346
	1.0 V/s	0.298	0.305	0.320	0.329
	10.0 V/s	0.284	0.290	0.303	0.311
	Av.	0.297	0.304	0.320	0.329
α (theor)	0.1 V/s	0.319	0.339	0.335	0.361
	1.0 V/s	0.307	0.327	0.318	0.344
	10.0 V/s	0.293	0.313	0.298	0.325
	Av.	0.306	0.326	0.317	0.343
α (exp[al])[g]	0.1 V/s	0.308		0.341	
	1.0 V/s	0.318		0.341	
	10.0 V/s	0.281		0.298	
	Av.	0.304		0.315	

Notes: [a]In DMF + 0.1 M *n*-Bu_4NBF_4, at a glassy carbon electrode at 25 °C.
[b]In V versus SCE.
[c]$D = 10^{-5}$ cm^2 s^{-1}.
[d]In eV.
[e]In Å.
[f]From literature thermochemical data (see text).
[g]The α value at each scan rate is derived from $E_{p/2}$–E_p, and the average also contains the value derived from $\partial E_p/\partial \log v$.

reduction potential, is in all cases very negative to the standard potential of the $RX + e^- \rightarrow R^\bullet + X^-$ reaction.

A more direct test of the variation of the transfer coefficient with the driving force would be to examine its possible variation with the scan rate in cyclic voltammetry. Such experiments[6] have indeed shown that such a trend exists. However, the variation of α with the driving force is detected with less precision than in the case of outer-sphere electron transfer (see Refs. 2s and 11 and references cited therein) because of the necessity of using as an electrode glassy carbon instead of mercury, limiting the range of scan rates in which meaningful experiments can be carried out. The detected variations are therefore not as distinctly beyond the experimental uncertainty as in the case of outer-sphere electron transfer reactions.

Table 3 shows the results obtained along the same lines of an even more detailed investigation of benzyl chloride and bromide in DMF,[12] where several scan rates instead of one (as in the preceding case) were used. We see that there is a good agreement between the predictions of the theory and the experimental values of the intrinsic barrier. The agreement is again better when potentials and the bond dissociation energies are estimated directly from literature RX and $R^\bullet$ gas-phase data than when using the second procedure ($\Delta G_{RX/R^\cdot} \neq 0$). The same is true for the values of the transfer coefficient. We see that there is a definite trend for the value of α to decrease at higher scan rates, and consequently the driving force, as predicted by the theory. This provides good evidence for the quadratic character of the activation–driving force relationship to be added to the quantitative agreement between the experimental and predicted value at each scan rate.

2.3. Application to Determination of Bond Dissociation Energies, Standard Potentials, and Intrinsic Barriers

Since the dissociative electron transfer theory predicts so well the experimental data for the reductive cleavage of unsubstituted benzyl halides where the standard potential and the bond dissociation energies are known, it may be used to estimate these two quantities when values are not available from independent sources. This approach has been used in the case of a series of arylmethyl halides according to the following procedure:

The standard potential of the $RX/R^\bullet + X^-$ couple, E^0_{RX}, may be decomposed as

$$E^0_{RX} = -D_{RX} - T(\overline{S}_{RX} - \overline{S}_{R^\cdot} - \overline{S}_{X^\cdot}) + E^0_{X^\cdot/X^-} \tag{27}$$

where D_{RX} is the C – X bond dissociation energy, $E^0_{X^\cdot/X^-}$ is the standard potential of the $X^\bullet/X^-$ couple, and $\overline{S}_{RX}$, $\overline{S}_{R^\cdot}$, $\overline{S}_{X^\cdot}$ are the partial molar entropies of the subscript species. When one compares, for each X, one RX in the series to $PhCH_2X$, $E^0_{X^\cdot/X^-}$ and $\overline{S}_{X^\cdot}$ do not change, whereas the variation in $\bar{S}_{RX} - \bar{S}_{R^\cdot}$ is likely to be negligibly small. Thus, to a good approximation,

$$E^0_{RX} = (E^0_{PhCH_2X} + D_{PhCh_2X}) - D_{RX}$$

For a given halogen, the following relationship between the standard potential and the bond dissociation energy applies for all compounds in the series:

$$E^0 = -D + C \tag{28}$$

C being obtained from the unsubstituted benzyl halides values

$$C = E^0_{PhCH_2X} + D_{PhCH_2X}$$

Treating the cyclic voltammograms along the same linearization procedure as just described, the value of D may thus be derived from the experimental value of the peak potential E_p as follows. The activation free energy at the peak potential, $\Delta G^{\neq}$, is obtained again from Eq. (18). Application of the activation–driving force relationship in Eq. (5) leads to

$$\Delta G^{\neq} = \frac{D + \lambda_0}{4}\left(1 + \frac{E_p - C - \phi_r + D}{D + \lambda_0}\right)^2 \tag{29}$$

and thus to

$$D = \tfrac{1}{2}\left\{[(\lambda_0 + E_p - C - \phi_r - \Delta G^{\neq})^2 + 4\lambda_0\Delta G^{\neq} - (\lambda_0 + E_p - C - \phi_r)^2]^{\frac{1}{2}} - (\lambda_0 + E_p - C - \phi_r - \Delta G^{\neq})\right\} \tag{30}$$

λ_0 is obtained from a by means of Eq. (24), a being estimated using Eqs. (22) and (25).

The values of D thus derived from the experimental E_p values are listed in Table 4. Rather than absolute values, the difference between the bond-dissociation energies of the considered compounds and those for the corresponding benzyl bromide or chloride are reported since the calculations have taken, through the value of C, as pivotal quantities the thermochemical data pertaining to these two compounds.

As seen from Table 4, there are significant differences in bond dissociation energies between the cyano- and carbomethoxy-benzyl bromides and benzyl bromide as well as between anthracenylmethyl chloride and benzyl chloride. Since the values of $Z(RT/F\nu D)^{1/2}$ and λ^0 do not vary much in the whole series, the essential reason that the

Table 4. Determination of the Bond-Dissociation Energy Increments from the Cyclic Voltammetric Peak Potential and Test of the Quadratic Character of the Activation–Driving Force Relationship[a] in the Reduction of Arylmethyl Halides

Compound	a_X (Å) / a_{RX} (Å) / a (Å) / λ_0 (eV)	C (eV) / $Z \times 10^{-3}$ (cm/s)	Scan rate (V/s)	$-E_p$ (V versus SCE)	$-\phi_R$ (mV)	$\Delta G^{\neq}$ (meV)	$-\Delta D$ (meV) [b]	$-\Delta D$ (meV) [c]	$-E^0$ (V versus SCE) [b]	$\Delta G_0^{\neq}$ (meV) [b]	α pred [b]	α exp[al] [d]
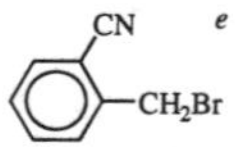 CN, CH₂Br [e]	1.96	1.665	0.1	1.44	72	341	156	180	0.479	717	0.345	0.367
	3.69		1.0	1.53	74	312	155	193	0.480	718	0.330	0.318
	2.88	4.63	10	1.63	76	283	151	207	0.484	718	0.314	0.289
	0.724		Av				154	193	0.481	718	0.330	0.315
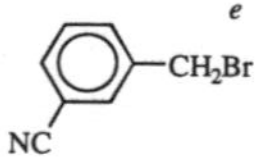 NC, CH₂Br [e]	1.96	1.665	0.1	1.62		341	50	60	0.584	744	0.339	0.341
	3.69		1.0	1.72		312	44	67	0.591	745	0.324	0.318
	2.88	4.63	10	1.83		283	34	73	0.601	747	0.308	0.265
	0.724		Av				43	67	0.592	745	0.324	0.293
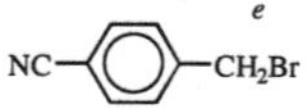 NC–, –CH₂Br [e]	1.96	1.665	0.1	1.46	72	341	149	166	0.491	719	0.344	0.341
	3.69		1.0	1.54	74	312	149	187	0.486	719	0.330	0.329
	2.88	4.63	10	1.65	77	283	139	193	0.496	721	0.313	0.298
	0.724		Av				149	182	0.491	720	0.329	0.332
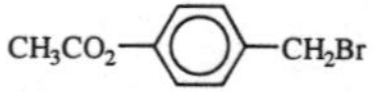 CH_3CO_2–, –CH₂Br	1.96	1.665	0.1									
	4.22		1.0									
	3.01	3.79	10	1.56	74	278	188	253	0.447	701	0.315	0.341
	0.693											
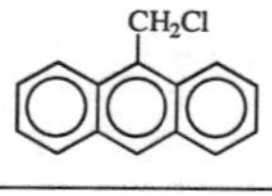 CH₂Cl	1.81	2.230	0.1	1.29	68	336	599	613	0.161	776	0.329	0.398
	4.22		1.0	1.36	70	307	612	627	0.149	773	0.315	0.382
	2.84	3.75	10	1.44	71	278	621	647	0.139	770	0.301	0.367
	0.714						611	629	0.150	773	0.315	0.372

Notes: [a]In DMF + 0.1 M n-Bu_4NBF_4, at a glassy carbon electrode at 25 °C.
[b]Rigorous calculation.
[c]Approximation.
[d]The indicated average value is the mean of the average of the values derived from $E_{p/2} - E_p$ and the value derived form $\partial E_p/\partial \log v$.
[e]In the presence of 50 mM benzoic acid.

reduction potential, at each scan rate, varies from one compound to the other derives from the variation of the bond-dissociation energy through the corresponding variations of E^0 and $\Delta G_0^{\neq}$.

The bond-dissociation energies of a series of 4-substituted benzyl bromides have been recently determined using photoacoustic calorimetry[13], a technique entirely independent of the procedures that led to the values gathered in Table 3. For 4-cyanobenzyl bromide there is a fair agreement between the photoacoustic calorimetric value (0.216 eV with a gap of 77 meV between the extreme values determined and the electrochemical value (0.149 eV), not far from the magnitude of the experimental uncertainty and taking into account that the photoacoustic calorimetric value was obtained in a different medium, a 3:1 diethylsilane–benzene mixture, less polar than DMF. The trend that the bond dissociation energies are weakened by an electron-withdrawing substituent on the benzene ring (assigned mostly to changes in the formation enthalpy of the starting bromide rather than of the radical[13]) is thus confirmed. As expected along these lines, the ΔD for the *ortho*-substituted cyano derivative is close to that of the *para*-substituted derivative and distinctly lower for the *meta*-substituted derivative. There is a much larger weakening of the bond when passing from benzyl to anthracenylmethyl chloride.

The difference in peak potentials may thus be utilized to estimate the difference between the bond-dissociation energies of two compounds bearing the same halogen and thus to derive the bond-dissociation energy of an unknown compound from that of a compound for which thermochemical data are available from other sources. The procedure described above may be made simpler and quicker by neglecting the differences in Z, D, and λ_0 for two compounds of similar size and the quadratic character of the activation–driving force relationship in the case where the peak potentials at the same scan rate are not too distant from one another. Under such conditions,

$$\Delta G^{\neq} = 0 \quad \text{and thus} \quad \Delta D = -\tfrac{2}{3}\,\Delta E_p \tag{31}$$

As seen from Table 4, the results of this quick estimation are not too different from those of the more rigorous procedure used previously.

The quadratic character of the activation–driving force relationship may nevertheless be tested in this series of compounds as accomplished previously with benzyl chloride and bromide. In this connection, Table 4 contains a comparison between the experimental values of the transfer coefficient α (exp[al]) and the values of α (pred) obtained by injecting into

Eq. (26) the values of $\Delta G_0^{\neq}$ derived from the experimental values of E_p. It is seen that the agreement is again satisfactory and that there is also a distinct trend for the experimental values to decrease on raising the scan rate (i.e., on increasing the driving force), as expected from the quadratic character of the activation–driving force relationship.

As reported in Table 4, the values of E^0 and $\Delta G_0^{\neq}$ may be predicted on the basis of the preceding estimate of D using Eqs. (28) and (7), respectively.

Another example of the determination of bond dissociation energies from cyclic voltammetric electrochemical data concerns the reductive cleavage of vicinal dibromides.[8a,c] It has been shown that the overall reaction leads to the formation of the corresponding olefin with elimination of the two bromides through the intermediacy of the β-bromo radical:

Br Br $+ e^-$ ⇌ • Br $+ Br^-$

formed upon dissociative injection of a first electron. The β-bromo radical is reduced along a second dissociative electron transfer at a potential higher than the reductive potential of the starting dibromide:

• Br $+ e^-$ ⇌ $+ Br^-$

The bond dissociation energies thus found are systematically lower than predicted on the basis of simple additivity rules in the whole series of 14 compounds investigated, with one exception; a compound (see below) where the two bromines are in vicinal equatorial positions on a sterically stabilized cyclohexane ring:

Br Br

3(e),4(e)-dibromo-1(e) methylcyclohexane

Br Br

3(a),4(a)-dibromo-1(e) methylcyclohexane

The difference in bond-dissociation energy with the axial–axial isomer (see above) is 4.8 Kcal/mol. This is a measure of the stabilization of the β-bromo radical resulting from electron delocalization over the C–C–Br framework in the axial–axial isomer by contrast to the equatorial–equatorial isomer, where this delocalization is prevented by the fixed unfavorable arrangement of the orbitals of the three atoms. This value can also be taken as an approximate measure of the stabilization of the β-bromo radical in the series. For the compounds that do not possess a constrained geometry of the two C–Br bonds, the ensuing decrease of the standard potential and the intrinsic barrier of the dissociative electron transfer allows one to conclude that the antiperiplanar conformer is the easiest to reduce and that the reduction of each compound thus proceeds through this conformer along a CE-type mechanism[1] (the chemical reaction C being here the conversion of the other conformers into the antiperiplanar conformer, while E designates the dissociative electron transfer by which the antiperiplanar conformer is reduced).

Along the same lines, the dissociative electron transfer theory has been recently used to unravel the intramolecular interactions in radicals resulting from the reductive cleavage of 1,3-dihaloadamantones, 2,3-dihalobicyclo(1.1.1.)pentanes, and 1,2-dihalobicyclo(2.2.2.)octanes.[14]

2.4. Homogeneous Dissociative Electron Transfer

The theory of dissociative electron transfer described above is expected to apply to homogeneous reactions as well as to electrochemical reactions. The dynamics of the reaction is then predicted to obey the quadratic activation–driving force relationships [Eq. (1)] and the intrinsic barrier to be the sum of:

- one fourth of the bond dissociation energy,
- a solvent reorganization term, and
- a term related to the changes in length and angles of the bonds not being broken in the acceptor as well as in the donor [Eq. (7)].

To test the theory, one must dispose of a series of electron donors that behave in an outer-sphere fashion *vis-à-vis* the electron acceptor that undergoes bond cleavage. Analysis of the problem would also be simpler if the changes in bond lengths and angles in the donor on electron transfer are negligible and if their involvement in solvent reorganization is small and about constant in the series.

The reactions of aromatic and heteroaromatic anion radicals with aliphatic halides in polar nonacidic solvents such as DMF have been investigated in this connection.[6,15] One attractive feature of these systems is that the anion radicals may be generated electrochemically and the kinetics of the reaction measured by means of the variations of the current flowing through the electrode in the framework of the usual electrochemical techniques such as cyclic voltammetry. How the characteristic rate constants may be extracted from the current responses and their variations with experimental parameters such as the concentrations of donor and acceptor and the scan rate has been described elsewhere (see Refs. 1 and 16 and references cited therein). Besides the possibility of measuring the kinetics by means of an electrochemical technique, the main advantage of the *in situ* electrochemical generation of the electron donor is that it need not be stable toward any species present in the

$$\text{(P)} + e^- \rightleftharpoons \text{(Q)}$$

$$\text{(Q)} + \text{RX} \xrightarrow{k} \text{(P)} + \text{R}\bullet + \text{X}^-$$

$$\text{(Q)} + \text{R}\bullet \xrightarrow{k_r} \text{(P)} + \text{R}^- \qquad (1')$$

$$\text{(Q)} + \text{R}\bullet \xrightarrow{k_a} \text{[H,R-cyclohexadienyl anion]}$$

$$\text{[H,R-cyclohexadienyl anion]} + \text{H}^+ \longrightarrow \text{[H,R-dihydrobenzene]}$$

$$\text{[H,R-cyclohexadienyl anion]} + \text{RX} \longrightarrow \text{[R,R-dihydrobenzene]} + \text{X}^-$$

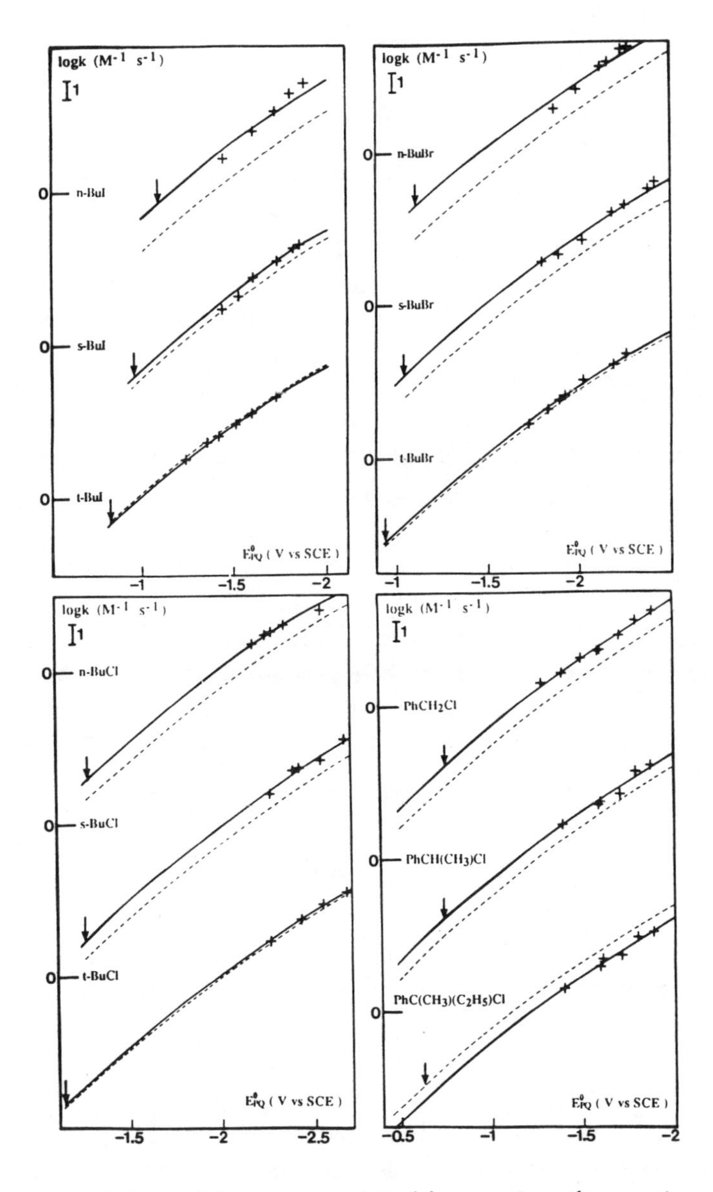

Figure 3. Variations of the rate constant of the reaction of aromatic anion radicals with alkyl halides (in DMF + 0.1 M n-Bu_4BF_4) with the standard potential of the mediator couple. + : experimental points. ↓ : location of the standard potential of the $RX/R^{\bullet}+X^{-}$ couple. Full lines : best fit to Eq. (1). Dashed lines : theoretical predictions.

medium that would be less reactive than the electron acceptor under investigation.

The experimental results obtained with *n*-, *s*-, and *t*-butyl iodides, bromides, and chlorides[6] are recalled in Figure 3. The cyclic voltammetric kinetic data were treated according to reaction scheme (1′), where the aromatic or heteroaromatic mediator is schematically represented by a benzene molecule. In most cases k_r/k_a is small and the reaction amounts to hydroalkylating the mediator. There are, however, cases where the reduction of the alkyl radical by the aromatic anion radicals is faster than their coupling, thus entailing a partial catalytic character of the cyclic voltammetric *P*/*Q* wave. In all cases, the dissociative electron transfer between the anion radical and the alkyl halides is rate determining. The analysis of the kinetic data thus leads to the values of the rate constant *k* of this step, as reported in Figure 3. The figure also shows the results of an analysis of the variations of the rate constant with the driving force of the reaction:

$$\Delta G^0 = E^0_{PQ} - E^0 \tag{32}$$

(where E^0 is the standard potential of the $RX + e^- \rightleftharpoons R^\bullet + X^-$ reaction and E^0_{PQ} that of the mediator reversible couple), similarly to what was done previously in the electrochemical case. Starting from the values of E^0, derived from thermochemical data as in the electrochemical case, a value of $\Delta G^{\neq}_0$ is derived from Eq. (1) (here $w_R = w_P = 0$) for each mediator by means of

$$\Delta G^{\neq}_0 = \frac{\left[(\Delta G^0 - 2\Delta G^{\neq})^2 - (\Delta G^0)^2\right]^{\frac{1}{2}} - (\Delta G^0 - 2\Delta G^{\neq})}{4} \tag{33}$$

where $\Delta G^{\neq}$, the activation free energy, is derived from the rate constant *k* according to

$$\Delta G^{\neq}\,(eV) = \frac{RT}{F} \ln\left(\frac{Z^{\text{hom}}}{k}\right) \tag{34}$$

Z^{hom}, the homogeneous collision frequency, is equal to $3 \times 10^{11}\ M^{-1}\ s^{-1}$ with good accuracy in the whole series of compounds. The mean value of $\Delta G^{\neq}_0$ over the whole series of mediators for each halide is reported in Table 5 under the heading $\Delta G^{\neq}_0$ (exp[al]) and was used to draw the full lines shown in Figure 3 using Eqs. (1) and (34). The good fit of this line with experimental rate data indicates the applicability of the activation–driving force relationship in Eq. (1) to all compounds.

The values of $\Delta G_0^{\neq}$ (exp^{al}) may now be compared with the theoretical values derived from the theory, which predicts that

$$\Delta G_0^{\neq} = \frac{\lambda_0^{\text{hom}} + D}{4} \tag{35}$$

According to Marcus[3], the solvent reorganization factor λ_0^{hom} can be obtained from

$$\lambda_0^{\text{hom}} = e_0^2 \left(\frac{1}{D_{\text{op}}} - \frac{1}{D_{\text{S}}} \right) \left(\frac{1}{2a_Q} + \frac{1}{2a} - \frac{1}{a_Q + a} \right) \tag{36}$$

(a_Q : hard-sphere radius of the mediator, 3.64 Å average in the whole series of mediators). a, the equivalent radius of the alkyl halides, can be estimated in the same way as already done in the electrochemical case (see Table 5). In fact, as shown by a previous ESR line-broadening investigation of self-exchange free energies of activation in an extended series of aromatic and heteroaromatic anion radicals,[7b] the Marcus expression [Eq. (36)], although predicting correctly the variation of the activation free energy with the hard-sphere radius, overestimates this variation by a factor on the order of 1.5. This observation may be related to an overestimation of the solvent reorganization free energy by the Marcus model and/or the neglect of the resonance energy in the estimation of the activation barrier, resonance energy that is anyway required to ensure the adiabaticity of the reaction and/or an underestimation of the pre-exponential factor. This point will be discussed in more detail later on. For the moment, it is assumed that the resonance energy and the pre-exponential factor are close in the present case to what they are in the electron exchange between aromatic anion radicals and parent hydrocarbons and thus that the solvent reorganization is estimated from the empirical relationship

$$\lambda_0^{\text{hom}}\ (\text{eV}) = 2.149 \left(\frac{1}{a_Q} + \frac{1}{a_{\text{RX}}} - \frac{2}{a_Q + a_{\text{RX}}} \right) \tag{37}$$

(where the hard-sphere radii are expressed in Å) following the same comparative strategy as in the electrochemical case. The results of these estimations are listed in Table 5. Using the values of the bond-dissociation energies derived from thermochemical data, one obtains the theoretical values of the intrinsic barrier free energies reported in the same table.

Table 5. Comparison between the Predictions of the Theory of Homogeneous Dissociative Electron Transfer and the Experimental Rate Data[a] for the Reaction of Butyl Halides[b] and Benzyl Chlorides[c]

Compound	$-E^0$ (*V versus SCE*)	a_{RX} (*Å*)	a (*Å*)	λ_0^{hom} (*eV*)	D (*eV*)	$\Delta G_0^{\neq}$ (*eV*) *theor.*	*exp*[al]
n-BuI	1.085	3.56	3.04	0.654	2.56	0.804	0.696
s-BuI[d]	0.946	3.58	3.58	0.653	2.38	0.758	0.736
t-BuI[d]	0.826	3.62	3.06	0.648	2.20	0.712	0.721
n-BuBr	1.19	3.50	2.82	0.687	3.00	0.922	0.827
s-BuBr[d]	1.049	3.50	2.82	0.687	2.95	0.909	0.844
t-BuBr[d]	0.929	3.54	2.83	0.687	2.87	0.888	0.884
n-BuCl	1.267	3.45	2.67	0.714	3.50	1.053	0.990
s-BuCl[d]	1.258	3.47	2.28	0.712	3.45	1.040	0.981
t-BuCl[d]	1.138	3.51	2.69	0.710	3.42	1.030	1.020
$PhCH_2Cl$	0.760	3.58	2.70	0.710	2.99	0.925	0.858
$PhCH(CH_3)Cl$	0.751	3.71	2.73	0.702	2.94	0.912	0.866
$PhC(CH_3)(C_2H_5)Cl$	0.631	3.84	2.77	1.96	2.91	0.900	0.945

Notes: [a]In DMF + 0.1 M *n*-Bu_4BF_4 at 20 °C unless otherwise stated.
[b]From Ref. 5.
[c]From Ref. 17.
[d]At 10 °C.

The same analysis repeated for a series of primary, secondary, and tertiary benzyl chlorides using the rate data reported in Ref. 16 leads to the results shown in Figure 3 and Table 5.

Overall, there is satisfactory agreement between the experimental and theoretical values of $\Delta G_0^{\neq}$ as seen in Table 5 and in Figure 3, where the ensuing reconstructed $\log k - E_{PQ}^0$ theoretical plots are represented as dashed lines.

It is clearly seen that the theory predicts quite correctly the differences observed on changing the halogen atom from I to Br and Cl, both at the level of the intrinsic barriers and of the slopes of the activation–driving force plots.

It distinctly appears, however, that, given the halogen and changing the nature of the reacting carbon, the agreement is better for the tertiary halides than for the secondary and primary halides. This trend appears even when account is taken of the fact that the uncertainties in the thermochemical data may affect the observed differences between $\Delta G_0^{\neq}$ (exp[al]) and $\Delta G_0^{\neq}$ (theor). From the mere fact that tertiary alkyl radicals are more stable than secondary and primary alkyl radicals it is expected that D increases in this series and E^0 becomes less and less negative. We thus expect that both $\Delta G_0^{\neq}$ (theor) and $\Delta G_0^{\neq}$ (exp^{al}) become larger and larger from tertiary to secondary and primary. This is what is observed in the first case, but the trend is opposite in the second. Thus taking the theory–experiment agreement as good for the tertiary halides, it is seen that the theory overestimates the intrinsic barrier in the case of secondary and primary halides even though the difference remains small (*circa* 100 meV at maximum).

These observations suggest that, in the absence of steric hindrance, the transition state would be stabilized, albeit by a small amount, *vis-à-vis* the transition state predicted for a dissociative electron transfer from an outer-sphere electron donor. The question thus arises whether this stabilization would involve bonded interactions. In favor of a positive answer is the previous finding that when reacting optically active 2-octyl iodide, bromide, or chloride with anthracene anion radical, although racemization prevails, a small but significant amount of inversion (on the order of 10%) is observed.[18] It may thus be envisaged that the outer-sphere dissociative electron transfer between the anion radical and the alkyl halide occurs along two competitive pathways, the outer-sphere dissociative electron transfer that would lead to racemization and an S_N2- type substitution leading to inversion:

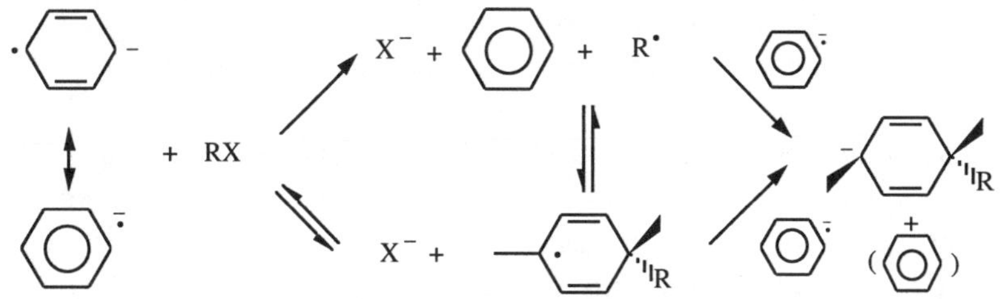

+ RX

X^- + + $R^{\bullet}$

X^- +

R

R

However, with stabilization energies on the transition state on the order of 50 to 100 meV as found here, resulting in an acceleration of the reaction of one to two orders of magnitude, one would expect a much larger ratio of inversion to racemization. It follows that a dissociation of the radical resulting from the S_N2 substitution into the aromatic hydrocarbon and the alkyl radical should also be envisaged. Thus, in the absence of strong steric hindrance, the main initial step would be an S_N2 substitution leading to a radical that would undergo a partial, though possibly preponderant, dissociation into the aromatic hydrocarbon and the alkyl radical in which the stereochemical integrity of the reacting carbon would be lost.

One consequence of the instability of the S_N2 substitution radical *vis-à-vis* the products of the outer-sphere dissociative electron transfer is that there is no driving force advantage of the former over the latter. One is thus led to conclude that the partially bonded transition state of the S_N2 pathway is stabilized *vis-à-vis* that of the outer-sphere electron transfer pathway by means of resonance between the two mesomeric structures :

$\cdot$⬡ $\bar{:}$ C $\cdot\cdot$ X ⟷ $\cdot$⬡ $\cdot\cdot$ C :X$^-$

to an extent able to overrun the slight driving force disadvantage and the fact that the entropy of the S_N2 transition state is expected to be smaller than that of the outer-sphere dissociative electron transfer transition state.

This latter point has been recently confirmed experimentally.[15c,d] The variation of the rate constant with temperature was indeed found to be different for *n*-BuBr as compared to *t*-BuBr and other sterically hindered alkyl bromides. Although the present level of experimental accuracy actually appears insufficient to detect the small change in slopes of Arrhenius plots that the competition between the ET and S_N2 pathway would predict, both series of experiments indicate that the average activation entropies are smaller with the primary bromide than with sterically hindered bromides (by ca. 10 eu).

There is thus convergent evidence, both kinetic (comparison of the activation–driving force plots, temperature dependence studies) and stereochemical that, in the absence of steric hindrance, the reaction of aromatic anion radicals follows, with a slight kinetic preference, an S_N2 rather than an outer-sphere dissociative ET pathway even though most of the stereochemical integrity of the reacting carbon is ultimately lost. The kinetic preference is, however, small and in practice does not support seriously the qualms recently expressed[15d] as to the use of aromatic anions radicals as standards for an outer-sphere dissociative electron transfer behaviors in the investigation of the mechanism by which other nucleophiles (electron donors) react with primary halides. In this connection the method of the "kinetic advantage" consists in locating the point corresponding to an unknown electron donor in the activation–driving force plot constructed from a series of aromatic anion radicals. The S_N2 character of the reaction of an unknown electron donor is then derived from the observation that its representative point is located significantly above the aromatic anion radical line. This approach was first proposed and applied in the case of electrochemically generated iron

(I) and cobalt (I) porphyrins[15a] and further applied to organic nucleophiles[15b,19,20] and to other low-valence metalloporphyrins.[15c] For nucleophiles that react with a primary or secondary halide faster than an aromatic anion radical of the same standard potential, such as unencumbered iron (I) or cobalt (I) porphyrins, it can be concluded *a fortiori* that an inner-sphere mechanism is followed. The ambiguity will affect only the cases where the unknown electron donor point is located in the close vicinity of the aromatic anion radical line. As an example it may be noted that sterically encumbered iron (I) porphyrins are located slightly below the line. At any rate, it should be emphasized that the location of the point in the close vicinity of the aromatic anion radical line that the S_N2 character of the initial step does not imply the stereoselectivity of the overall reaction.

Coming back to the question of the activation entropies in outer-sphere ET and S_N2 pathways, the above discussion seems to contrast with recent *ab initio* calculations concerning the reaction of the cation radical of ethane with hydrogen sulfide:[21a]

$$H_2S + C_2H_6^{\cdot +} \xrightarrow{ET} H_2S^{\cdot +} + C_2H_6$$

$$H_2S + C_2H_6^{\cdot +} \xrightarrow{S_N2} H_2SCH_3^{+} + CH_3^{\cdot}$$

The situation is in fact rather different from that of the preceding discussion since an electron transfer not involving any bond breaking is compared to a reaction where concerted bond formation and bond breaking occur. The decrease in activation entropy on passing from S_N2 to ET is then essentially related to the fact that the C–C bond is broken in the former case and not in the latter. On the other hand, in the preceding discussion, the C–X bond is broken in both reactions, and in addition, a new bond is formed concertedly in the S_N2 case.

Let us come back to the question of the magnitude of the resonance energy in the transition state of outer-sphere dissociative electron transfer reactions. It has recently been suggested[21b] that, in such processes, the resonance energy is much larger than previously thought and that the reaction should therefore be better described as an inner-sphere rather than an outer-sphere electron transfer even though it does not involve the concerted formation of a new bond. This conception was even extended to self-exchange electron transfer between aromatic anion radicals so far viewed as purely outer-sphere electron transfer reactions. The argument that was used to gain evidence in favor of these assertions from the experimental data calls for the following remarks.

Concerning the homogeneous self-exchange reaction between aromatic anion radicals and the parent hydrocarbons, the experimental fact,[7b] already noted above, that the strict application of the Marcus model of solvent reorganization in its primitive version overestimates the observed activation barrier is taken as an indication of the importance of the resonance energy. This is indeed estimated as the difference between the experimental and Marcus predicted barrier leading to an average of 2.5 Kcal/mol (recalculation of the difference from the original data,[7b] points to a slightly lower value, 1.8 Kcal/mol). That this difference can be attributed to the resonance energy exclusively is not a completely settled question. The Born model of solvation that is used in the Marcus model is known to overestimate the solvation free energies and to ignore the specific Lewis acid–base interaction between solvent and solute (see Ref. 5i and references cited therein). However, this may be compensated for by solvent dynamic effects that were not taken into account in the abovementioned discussion of aromatic anion radicals. Such a compensation has recently been shown to take place the cobalticinium/cobaltocene couple.[5i] On the other hand, as discussed earlier, the pre-exponential term may be larger than the bimolecular collision frequency (3×10^{11} M^{-1} s^{-1}) used in the preceding analyses, leading to an underestimate of the experimental reorganization barrier and thus a smaller value for what is attributed to the resonance energy. Some resonance energy in the transition state is required to ensure the adiabaticity of the electron transfer reaction (i.e., that the electronic transmission coefficient κ_{el} be close to 1):[5a,b]

$$\kappa_{el} = \frac{2[1 - \exp(-\nu_{el}/2\nu_n)]}{2 - \exp(-\nu_{el}/2\nu_n)} \tag{38}$$

ν_n being the nuclear frequency and ν_{el} the electronic frequency that can be estimated, in the high temperature limit, from

$$\nu_{el} = \frac{H^2}{h}\left(\frac{\pi^3}{4\,RT\,\Delta G_0^{\neq}}\right)^{\frac{1}{2}} \tag{39}$$

where H is the electronic coupling matrix element of the precursor and successor complexes. In the present case, it follows that a resonance energy of 1 kcal is largely sufficient to ensure the adiabaticity of the reaction. Recent HMO estimation of the resonance energy in the adiabatic self-exchange reaction of TCNQ and TTF led to values of 1.3 and 2.1 kcal/mol, respectively,[5h] not far from the average value of 1.8 kcal/mol found from the difference between the application of Marcus estimation of λ_0 including a crude estimation of the pre-exponential factor and the experimental data. Such a value thus appears a reasonable estimate of the resonance energy even if it most probably results from compensation of various small systematic errors.

Application of the same strategy to the dissociative electron transfer between aromatic anion radicals to butyl halides has led to the suggestion[21b] that the resonance energy is even larger (4 kcal/mol) when the data pertaining to primary, secondary, and tertiary iodides, bromides, and chlorides are averaged. In fact, as shown earlier, with the tertiary halides, there is an excellent fit between experiment and the theoretical predictions using the experimental values of λ_0^{hom} derived from the aromatic anion radical self-exchange data. This means that the resonance energy embedded in these data is simply carried over to the tertiary alkyl halide case and thus that the resonance energy is approximately the same. Larger resonance energies, up to 4 kcal/mol, are found for the primary and secondary halides as a result of the (at least partial) inner-sphere (S_N2) character of the reaction falling in line with the presence of inversion compounds among the reaction products and with the slightly more negative activation entropies then observed.

The other strategy used in Reference 21b to estimate the resonance energy in the homogeneous self-exchange reaction of aromatic anion radical as well as in their reaction with butyl halides was to compare the electrochemical and homogeneous data analyzing both of them though the Marcus model assuming that

$$\Delta G^{\neq}_{0,\text{hom}} = 2\,\Delta G^{\neq}_{0,\text{electrochem}}$$

and that resonance energies are negligible in the electrochemical case. Resonance energies as large as 6 and 10 kcal/mol were found in this way for the self-exchange and alkyl halide reactions, respectively. It should be noted first that these values are overestimated by ca. 2.5 kcal/mol by the simple fact that the homogeneous and heterogeneous collision frequencies are taken as equal to 10^{11} M^{-1} s^{-1} and 10^4 cm s^{-1}, respectively, instead of the 3×10^{11} and 4×10^3 cm s^{-1} derived from the gas kinetic theory. Discrepancies between the homogeneous and heterogeneous barriers in the framework of the Marcus model have been noted previously and assigned to image force effects and work terms (double layer correction).[22] The Marcus model underestimates the solvent reorganization energy because image force effects are calculated under the assumption that the hard-sphere that approximated the reactant lies against the electrode, whereas it is more likely to be located in the outer Helmoltz plane (i.e., farther from the electrode's surface). In addition, the presence of the solvent and of a large concentration of the supporting electrolyte cations may well shield partially the electrode and further decrease the image force effects, leading to an estimate of the solvent reorganization energy closer to the predictions of the Hush model, in which these effects are completely neglected. As discussed earlier, the actual situation is most probably more complex since other effects affecting the pre-exponential factor (such as solvent dynamics and effects resulting from the influence of the large concentration of supporting electrolyte cations at the reaction site on the medium reorganization energy) are likely to interfere. One has also to envisage the possibility that some resonance energy may well exist in the electrochemical case as well (some theoretical approaches even predict that it should be larger than in the homogeneous case[23]). It follows that the strategy of comparing electrochemical with homogeneous rate data has little chance to produce reliable estimates of the resonance energy in the homogeneous case in the present state of homogeneous and heterogeneous electron transfer theory.

It may be concluded that, in practice, the approach described earlier that consists in assuming that the resonance energies and pre-exponential factors are approximately the same and that the solvent reorganization free energies follow the same proportionality to the inverse of the hard-sphere radius in the self-exchange reaction of aromatic radicals and in their reaction with alkyl halides leads to a satisfactory rationalization of the experimental data in the case of sufficient steric hindrance. In particular, it allows one to interpret in terms of an increase of the resonance energy the small differences found among the kinetics of primary, secondary, and tertiary halides, in accord with the entropy and stereochemical data.

The reaction of aromatic anion radicals (including copper, zinc, and free-base porphyrins) with the same vicinal dibromides that have been studied electrochemically as described earlier has also been investigated.[8a,c] The application of the same strategy for minimizing errors in the estimation of the pre-exponential factor and of the solvent reorganization energy led to an excellent consistency of the homogeneous and electrochemical barriers. The antiplanar preference already observed in electrochemistry is confirmed by the analysis of the homogeneous data.

2.5. Symmetry Factor (Transfer Coefficient, Brönsted Slope, Tafel Slope)

If the activation–driving force relationship is quadratic, as depicted in Eq. (1), the symmetry factor α (the transfer coefficient or Tafel slope of the electrochemist, or the Brönsted slope of the homogeneous chemist) is expected to decrease upon raising the driving force in a linear way [Eq. (4)]. If these predictions are correct, one might have a powerful experimental tool for analyzing mechanisms by use of the fact that a small α (i.e., distinctly below 0.5) would then indicate that the system reacts under a strong driving force and *vice versa*.

It is thus important to test this aspect of electron transfer dynamics model for both outer-sphere and dissociative electron transfers. This is easier in the electrochemical case than in the homogeneous case for several reasons. One is that the driving force may be changed in a precise and continuous manner by simply varying the electrode potential, whereas in the homogeneous case changes in driving force require the use of a series of counterreactants, which each introduce their own internal and external reorganization factors into the activation–driving force relationship together with those of the compound with which they exchange one electron. It is true that the work terms (double-layer correction) are more difficult to estimate in the electrochemical case, but they do not vary very much within the available range of driving forces.

In the case of outer-sphere electron transfer to aromatic molecules, the variation of α with the electrode potential (i.e., of the symmetry factor with the driving force) has received backing from unambiguous experimental evidence and found to be of the same order of magnitude as predicted by the Marcus–Hush model (see Ref. 11 and Refs. cited therein). In the case of dissociative electrochemical electron transfer to alkyl[6] and benzyl halides,[12] there is also evidence that the transfer coefficient decreases on raising the driving force. The increases in the driving force as obtained by raising the cyclic voltammetric values (100 $V\ s^{-1}$) are not so large as to fear the interference of residual ohmic drop effects after positive feedback compensation.[1]

Besides the variations of α with electrode potential that are necessarily small in accordance with Eq. (4), one is also struck by its small average value (ca. 0.3, i.e., distinctly smaller than 0.5). There is so far no other model that could predict such low values resulting from another cause than the fact that the reduction potential is quite negative with respect to the standard potential of the $RX/R^{\bullet} + X^-$ couple.

Detection of a contingent variation of the symmetry factor of homogeneous reactions with the driving force is a much harder task for the reasons given above, also taking account of the experimental uncertainty in rate measurements. There is not so far, even in the case of purely outer-sphere electron transfers, clear-cut direct evidence telling whether the symmetry factor does or does not vary with potential. The situation is the same for the dissociative electron transfer involving aromatic anion radicals and alkyl halides, in spite of recent claims that the $\log k - E^0_{PQ}$ plots are linear rather than (slightly) parabolic and that the theory of dissociative electron transfer should be amended accordingly.[15d,24] The most interesting recent results in this connection concern the reduction of *t*-BuBr by an extended series of aromatic anion radicals in DMF[15d] since, as discussed earlier, these electron donors behave in an outer-sphere fashion toward *t*-BuBr. The cyclic voltammetric data were extended toward lower rate constants by use of homogeneous kinetic techniques and toward larger rate constants by means of pulse radiolysis. Concerning this first extension, in view of the rather large scatter of the experimental points (when the dissociative electron transfer is slow it is particularly difficult to avoid competing reactions of the aromatic anion radical with electrophilic and reducible impurities), there is no real gain in the possibility of discriminating between the linear and slightly parabolic behavior (this can be easily simulated using the E^0 and $\Delta G_0^{\neq}$ values reported in Table 5 or values close to these to account for small differences in experimental conditions). As for the pulse radiolysis experiments, the absence of experimental details prevents the possibility of reproducing the data and evaluating the contingent sources of uncertainties and systematic errors. From another pulse radiolytic study,[24] it appears that for the same aromatic anion radical, the rate constant appears to be significantly larger than the electrochemical rate constant (by a factor of about 3). There might indeed be a systematic difference between the rate constants gained by one or the other methods arising from different experimental conditions (for example, the pulse radiolytic experiments are usually carried out in the absence of supporting electrolyte and at much lower concentrations than in electrochemistry). On the other hand, rate constants close to the diffusion limit do not allow a precise determination of the activation-controlled rate parameters. Taking these uncertainties into account, it appears that the available data neither disprove nor prove for the time being the slight variation of α with the driving force expected from the theory. What is, however, clear in these experiments is that the average value of α one can derive from a straight-line fitting, 0.4, is clearly below 0.5, a result which falls in line with what is expected from the theory taking into account the fact that most of the experimental points are located at potentials negative to the standard potential of the $RX/R^{\bullet} + X^-$ couple.

Adding to this that the electrochemical α is smaller than the average homogeneous α, in agreement with the larger driving force in the electrochemical case, it may be concluded that the quadratic model correctly predicts the magnitude of the symmetric factor, particularly the expectation that it should be distinctly smaller than 0.5 under large driving forces.

3. STEPWISE AND CONCERTED MECHANISMS: DISCRIMINATION AND CONTROLLING FACTORS

3.1. Discrimination Criteria in Electrochemical Reactions

As long as the lifetime of the anion radical of the starting compound

$$RX + e^- \underset{k_{-E}}{\overset{k_E}{\rightleftharpoons}} RX^{\bullet -}$$

$$RX^{\bullet -} \xrightarrow{k_d} R^{\bullet} + X^-$$

is long enough to permit detection by electrochemical kinetic techniques, the recognition of the stepwise character of the mechanism is rather straightforward. For example, in cyclic voltammetry, detection consists of the observation of the reoxidation wave of the anion radical on increasing the scan rate, that is, the observation of a chemically (albeit not necessarily electrochemically) reversible wave. The lower limit of detection falls, under favorable circumstances, in the submicrosecond range thanks to the use of ultramicroelectrodes.[16]

Discrimination between the two mechanisms is more difficult in cases where the intermediate $RX^{\bullet -}$ is so unstable that the cyclic voltammogram remains chemically irreversible over the whole range of accessible scan rates. How one can know whether this irreversibility in the reflection of a fast reaction following an outer-sphere electron transfer step or the reflection of the concertation of the two steps?

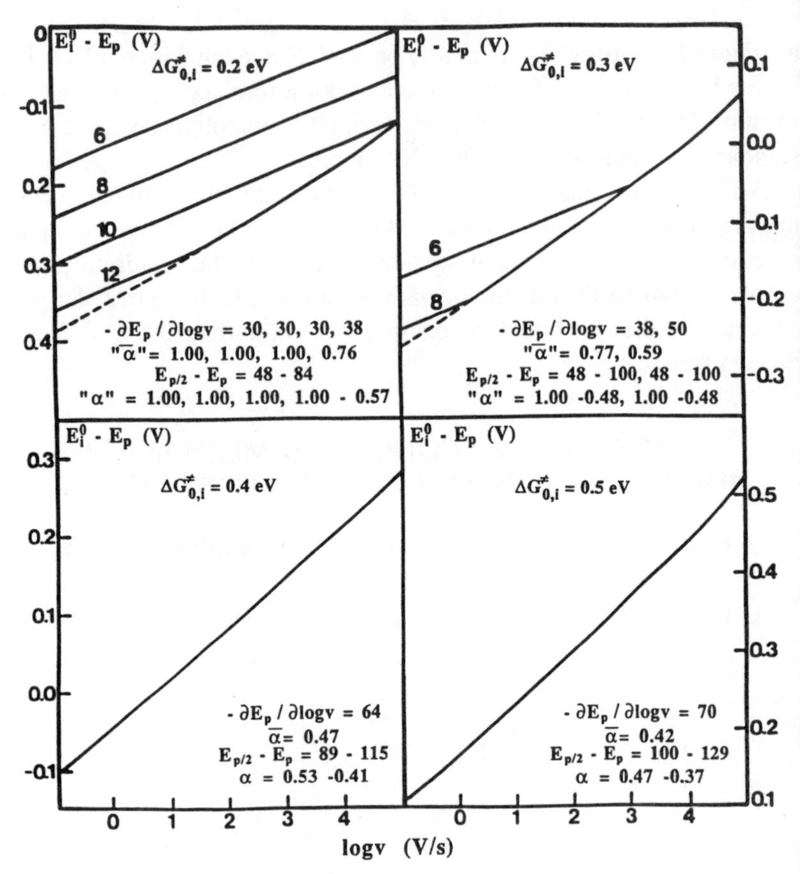

Figure 4. Stepwise mechanism in electrochemical reductive cleavages. Variations of the difference between the standard E_i^0 potential for the formation of the intermediate and the cathodic peak potential E_p with the scan rate for various values of the intrinsic barrier $\Delta G^{\neq}_{0,i}$ of the electron transfer and various values of the rate constant from the decay of the intermediate between 10^6 and 10^{12} s^{-1}. The successive values of $-\partial E_p/\partial \log v$(mV), "$\overline{\alpha}$" or $\overline{\alpha}$, $E_{p/2} - E_p$ (mV), and "α" or α correspond to the succession of the $(E_i^0 - E_p) - \log v$ plots from top to bottom. Temp. = 25 °C; $Z = 4.10^3$ cm s^{-1}; $D = 10^{-5}$ cm^2 s^{-1}.

In the first of these two cases, the anticipated cyclic voltammetric behaviors are those represented in Figure 4 concerning the variation of the cathodic peak potential E_p with the scan rate v and the peak width $E_{p/2} - E_p$. Rather than an opaque representation of the results of a general analysis, we have selected four typical values of the intrinsic barrier free energy $\Delta G^{\neq}_{0,i}$ (0.2, 0.3, 0.4, 0.5 eV) and four typical values of the rate constant of the follow-up reaction ($\log k_d = 6, 8, 10, 12$). The temperature is taken as 25 °C, and the potential difference ϕ_r between reaction site and solution is neglected. A value of the factor $Z\,(RT/D)^{1/2}$ is also needed in the computation. For this, two typical values, $Z = 4.10^3$ cm s^{-1} and $D = 10^{-5}$ cm^2 s^{-1} were taken. Figure 4 thus shows the variation of the potential difference $E_i^0 - E_p$ with $\log v$. E_i^0 is the standard potential of the formation of the intermediate $RX^{\bullet -}$, as diagrammed above, whereas we keep the notation E^0 for the standard potential of the dissociative electron transfer couple $RX/R^{\bullet} + X^-$. The results pertaining to other conditions are readily obtained from an adaptation of the analyses contained in Refs. 1 and 25, where an unconditionally linear activation–driving force relationship is used to describe the kinetics of electron transfer (Buttler–Volmer law) to the approach used in the preceding sections, where the quadratic activation–driving force relationship is linearized over each individual voltammogram but not when varying the scan rate [the pertinent equations are then Eqs. (15) and (18)]. In the case where ϕ_r is different from zero, it suffices to read the $(E_i^0 - E_p) - \log v$ curve concerning electron transfer as an $(E_i^0 + \phi_r - E_p) - \log v$ curve or, equivalently, to shift $\Delta G^{\neq}_{0,i}$ to $\Delta G^{\neq}_{0,i} - \overline{\alpha}\,\overline{\phi}_r$, where $\overline{\alpha}$ is the average transfer coefficient and $\overline{\phi}_r$ the average value of ϕ_r in the potential range of interest.

There is a competition between the reverse electron transfer and the cleavage reaction (associated with diffusion from the electrode) for the kinetic control of the overall reaction. This appears particularly clearly for the two smallest values of $\Delta G^{\neq}_{0,i}$ (i.e., the fastest forward and reverse electron transfers). At low scan rates, kinetic control is obtained by the follow-up cleavage. It shifts to forward electron transfer as the scan rate increases provided the reverse electron transfer is not too fast and the follow-up reaction not too slow; otherwise, the latter reaction is rate controlling over the whole range of scan rates. For the two largest values of the electron transfer intrinsic barrier, the reverse electron transfer is too slow to compete with the follow-up cleavage even at the lowest value of the rate constant of the latter (10^6 s^{-1}).

In each panel of Figure 4 we have reported the mean value of the slope of the $(E_i^0 - E_p) - \log v$ plot, $-\partial E_p/\partial \log v$, and the extreme values of the peak width over the whole range of scan rates. Average and bracketing values of α derive from[1,25]

$$\overline{\alpha} = (RT/2F)\ln 10/(\overline{-\partial E_p/\partial \log v}) \qquad (40)$$

and

$$\alpha = \frac{1.85\ RT/F}{E_{p/2} - E_p} \tag{41}$$

In the cases where kinetic control is achieved, at least partially, by the follow-up cleavage, we have put α between quotation marks to make it clear that we then do not deal with a true transfer coefficient (symmetry factor) pertaining to an electron transfer reaction but to an apparent transfer coefficient merely reflecting the slope of the E_p-log$v$ plot or the value of the peak width. In the first two sets of plots in Figure 4 ($\Delta G^{\neq}_{0,i}$ = 0.2, 0.3 eV), only the two limiting behaviors—control by the forward electron transfer and control by the follow-up cleavage—are represented. In fact, the passage between the two limiting behaviors covers a large portion of the range of scan rates.[25] This is taken into account in the mean values reported in Figure 4.

The main features that serve as diagnostic criteria for distinguishing the stepwise from the concerted mechanism are those pertaining to the relative location of the peak and standard potentials and to the value of α (or "α").

Concerning the first of these, it is important to remark that the standard potential for the generation of the intermediate E^0_i is positive relative to the peak potential in most cases, and when it may become negative to the peak potential (i.e., when electron transfer is fast), the gap is not very large, 0.4 V at maximum. The location of the peak potential may thus serve to estimate that of the standard potential in the framework of *a fortiori* reasoning as illustrated below.

The valuc of α (or "α") is also a useful criterion in the sense that it is contained between 1 and 0.42 over the whole range of values of $\Delta G^{\neq}_{0,i}$ and log k and has a value above or equal to 0.5 if $\Delta G^{\neq}_{0,i} \leq 0.3$ eV. As discussed later on, combination of the two types of criteria may also prove worthwhile.

The variations of the peak potential with scan rate and the peak width characterizing a dissociative electron transfer reaction are summarized in Figure 5. Since a bond is broken during the reaction, the intrinsic barriers are likely to be larger than is the case of an outer-sphere electron transfer. This is the reason that larger values of $\Delta G^{\neq}_0$, ranging from 0.5 to 1.2 eV, were selected in the representation given in Figure 5.

It is thus seen that in all cases the standard potential of the $RX/R^{\bullet} + R^-$ couple is positive to the cathodic peak potential and that the transfer coefficient is clearly below 0.5. The more it is so, the larger the intrinsic barrier. Discrimination between the stepwise and concerted pathways

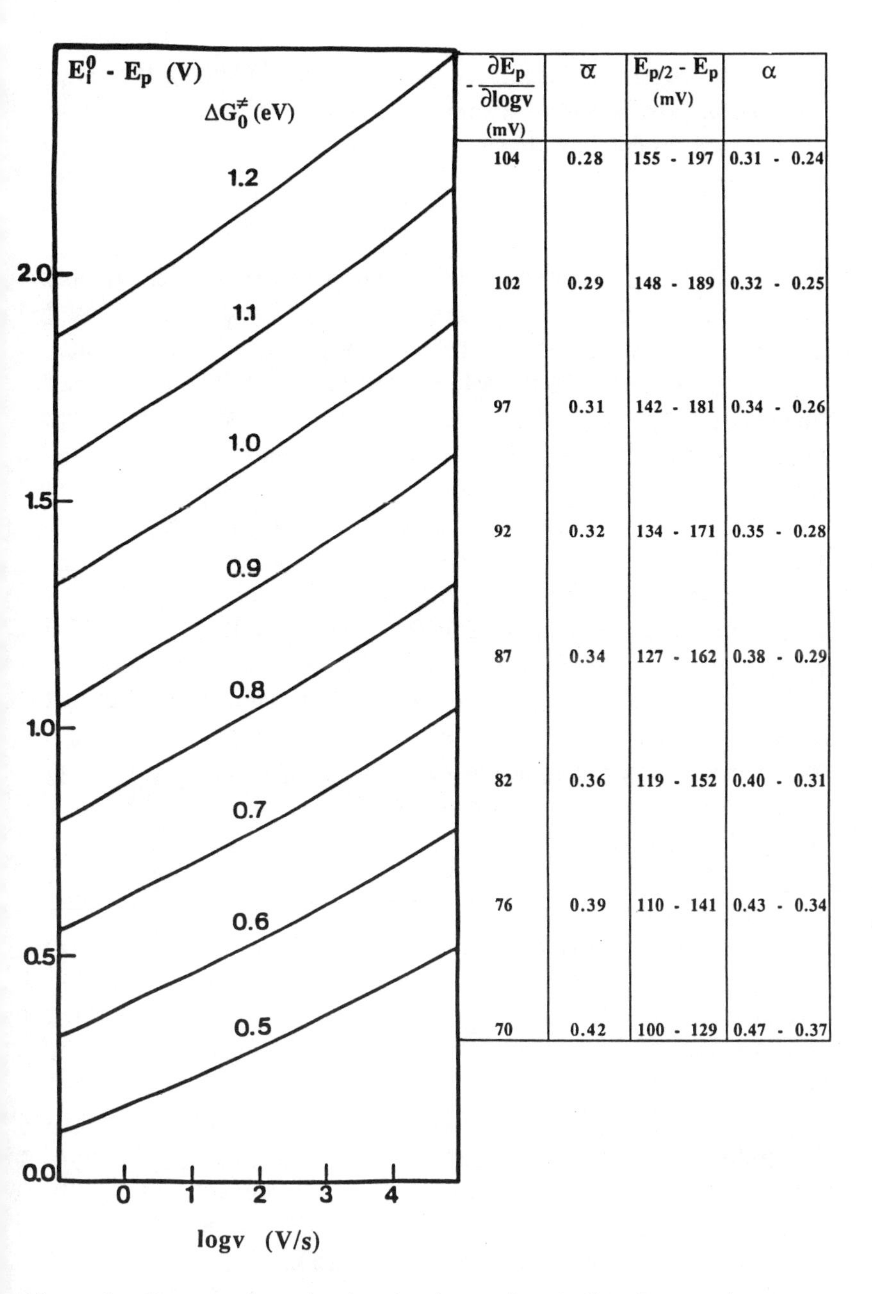

$-\frac{\partial E_p}{\partial \log v}$ (mV)	$\bar{\alpha}$	$E_{p/2} - E_p$ (mV)	α
104	0.28	155 - 197	0.31 - 0.24
102	0.29	148 - 189	0.32 - 0.25
97	0.31	142 - 181	0.34 - 0.26
92	0.32	134 - 171	0.35 - 0.28
87	0.34	127 - 162	0.38 - 0.29
82	0.36	119 - 152	0.40 - 0.31
76	0.39	110 - 141	0.43 - 0.34
70	0.42	100 - 129	0.47 - 0.37

Figure 5. Concerted mechanism in electrochemical reductive cleavages. Variation of the difference between the standard potential of the concerted reductive cleavage E^0 and the cathodic peak potential E_p with the scan rate for various values of the intrinsic barrier $\Delta G_0^{\neq}$. Temp.= 25 °C; $Z = 4.10^3$ cm s^{-1}; $D = 10^{-5}$ cm^2 s^{-1}.

may therefore use the location of the peak potential and/or the value of the transfer coefficient as diagnostic criteria, as illustrated below with several experimental examples.

3.2. Electrochemical Examples

Among the various studies in the field, the most striking example is that of the electrochemical reduction of benzyl and other arylmethyl halides in acetonitrile and DMF because one passes from one mechanism to the other according to the nature of the aryl group[10]. The nitrosubstituted benzyl chlorides and bromides undergo a stepwise electron transfer–bond breaking reaction whereas the other compounds in the series (cyanobenzyl, carbomethoxy benzyl, anthracenylmethyl, unsubstituted benzyl chlorides and/or bromides) undergo dissociative electron transfer.

All compounds in the series exhibit a first irreversible (at least at low scan rate) wave, at which the reductive cleavage of the carbon–halogen bond takes place, followed by a reversible wave corresponding to the reduction of the hydrogenolysis methyl compound or to that of the CH_2-CH_2-dimer into their respective anion radicals and dianions. 3-nitrobenzyl chloride is the most evident example of a stepwise mechanism since its first wave becomes reversible upon raising the scan rate up to 100 V s^{-1}. At low scan rate (between 0.1 and 2V s^{-1}), the wave is irreversible and the cathodic potential is a linear function of the log of the scan rate with a slope of –28 mV and a peak width of $E_{p/2} - E_p = 50$ mV, as expected from a rate-determining first-order cleavage following a fast outer-sphere electron transfer step (Figure 4 with small $\Delta G^{\neq}_{0,i}$). With 2- and 4-nitrobenzyl chloride there is a mixed kinetic control by electron transfer and cleavage: $\partial E_p/\partial \log v = 45$ and 38 mV and $E_{p/2} - E_p = 70$ and 60 mV, respectively (Figure 4).

These results indicate that the cleavage is faster than with the 3-nitrobenzyl chloride and somewhat faster for the *ortho* than for the *para* derivative. With the latter compound, chemical reversibility is reached at 300,000 V s^{-1}. At a lower scan rate, $\partial E_p/\partial \log v = -45$ mV and $E_{p/2} - E_p = 65$ mV, pointing again to a mixed kinetic control of electron transfer and cleavage.

Thus, for all for nitrobenzyl derivatives, the stepwise mechanism unambiguously takes place. In the case of 3- and 4-nitrobenzyl chlorides, the lifetimes of the anion radical could be determined: 12.5 ms and 0.25 μs, respectively. Another interesting result is that the standard potential

Table 6. Electrochemical Reduction of Arylmethyl Halides[a]. Transfer Coefficient Distance between Peak Potential and Standard Potential of the RX/RX$^{\bullet-}$ Couple

Compound	α^b	$E_i^0 - E_p$ [c] *(mV)*
C_6H_5–CH_2Cl	0.30	≥ 600
C_6H_5–CH_2Br	0.31	≥ 900
9-(chloromethyl)anthracene (CH_2Cl)	0.36	590
CH_3CO_2–C_6H_4–CH_2Br	0.34	850
o-CN–C_6H_4–CH_2Br	0.31	630
m-NC–C_6H_4–CH_2Br	0.29	700
NC–C_6H_4–CH_2Br	0.32	900

Notes: [a]In DMF + 0.1 M *n*-Bu_4BF_4, at a glassy carbon electrode at 25 °C.
[b]Average of the values derived from $\partial E_p/\partial \log v$ and $E_{p/2} - E_p$ between 0.1 and 10 V/s.
[c]E_p is the vlaue of the cathodic peal potential at 0.1 V/s.

and standard rate constant of the RX/R$^{\bullet-}$ could be measured in both cases. The apparent standard rate constant is on the order of 1 cm s^{-1}, corresponding to $\Delta G^{\neq}_{0,i} \cong 0.2$ eV. The standard potentials E_i^0 were found to be close (not more than 50 mV positive) to the standard potential of the corresponding nitrotoluenes, showing that the electron-withdrawing inductive effect of the CH_2Cl group *vis-à-vis* that of the CH_3 group is very small.

Passing now to the other members of the series, it is found that, for all of them, the transfer coefficient derived from $\partial E_p/\partial \log v$ or $E_{p/2} - E_p$ are much smaller than 0.5 (Table 6). At the same time, E_p is very positive relative to the standard potential of the corresponding toluene and therefore to the standard potential E_i^0 of the RX/R$^{\bullet-}$ couple (Table 6). By reference to Figure 4 ($\Delta G^{\neq}_{0,i} = 0.2$ eV), it is seen that the experimental difference $E_i^0 - E_p$ is much too large to fit with a stepwise mechanism. The low experimental values of α are also incompatible with this mechanism. Thus both the criterion based on the location of the cathodic wave and that based on the value of the transfer coefficient lead to the same answer, namely electron transfer and bond breaking are concerted for the whole series of compounds.

The behavior of aromatic halides[26] is strikingly different in the sense that in an extended series of aryl chlorides and bromides, and even for iodobenzene, the characteristics of the cyclic voltammetric cathodic peaks indicate that the reductive cleavage follows the stepwise mechanism (Table 7). For a large number of compounds, the cyclic voltammogram becomes reversible below 2000 V s^{-1}, showing unambiguously the intermediacy of the anion radical. For the other compounds in the series, where the wave remains irreversible up to these scan rates, the value of "α" shows either a complete kinetic control by the cleavage reaction, or a mixed kinetic control by the electron transfer and the cleavage reaction or by the initial electron transfer with a reduction potential close to the standard potential of the $RX/R^{\bullet -}$ couple (α = 0.5). For chloro- and bromobenzene, for example, which are expected to give rise to the fastest-cleaving anion radicals, the peak potential at 0.1 V s^{-1} is positive with respect to the $E^0_{RX/RX^{\bullet -}}$ only by 200 and 10 mV, respectively, whereas they are negative with respect to the standard potential of the $PhX/Ph^{\bullet} + X^-$ couple by 780 and 840 mV, respectively. In the case of PhI, the cathodic peak potential at 0.1 V s^{-1} is negative with respect to the standard potential of the $PhI/Ph^{\bullet} + I^-$ couple by 530 mV. Thus, even for the compounds that are expected to give rise to the most unstable anion radicals in the series, the reductive cleavage appears to follow the stepwise mechanism.

The electrochemical reduction of simple aliphatic halides such as the butyl iodides and bromides in DMF that have been mentioned in a preceding section again shows transfer coefficients that are much lower than 0.5 (Table 6). Unlike the situation in the preceding cases, there is no direct way of estimating the standard potential of the $RX/R^{\bullet -}$ couple if the reaction were proceeding through $RX^{\bullet -}$. However, the peak potentials at 0.1 V s^{-1} are located in the close vicinity of the standard potentials of phenyl iodide and bromide, respectively. The latter are certainly very positive with respect to the standard potential of the corresponding $RX/R^{\bullet -}$ couple in the aliphatic series. It follows that the reduction potentials are certainly very positive with respect to the standard poten-

Table 7. Electrochemical Reduction of Aryl Halides[a]

Aryl chlorides	$-E_i^0$ *(V versus SCE)*	"α" [b]
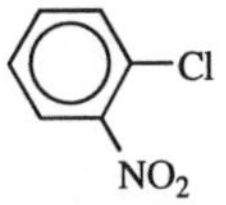 Cl, NO_2	0.99	reversible at $v \leq 2000$ V/s
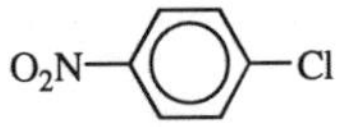 O_2N–, –Cl	1.05	reversible at $v \leq 2000$ V/s
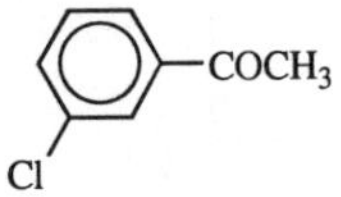 $COCH_3$, Cl	1.83	reversible at $v \leq 2000$ V/s
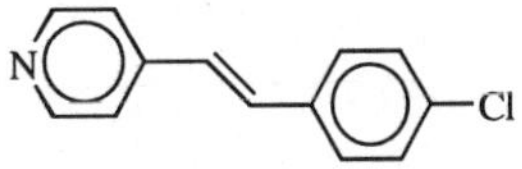 N, Cl	1.84	reversible at $v \leq 2000$ V/s
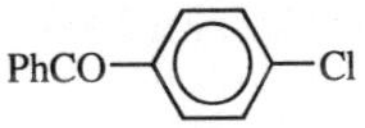 PhCO–, –Cl	1.64	reversible at $v \leq 2000$ V/s
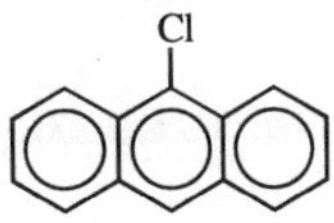 Cl	1.71	reversible at $v \leq 2000$ V/s
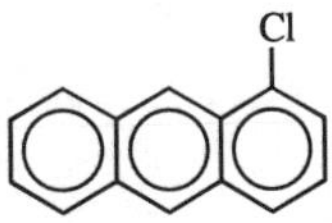 Cl	1.73	reversible at $v \leq 2000$ V/s
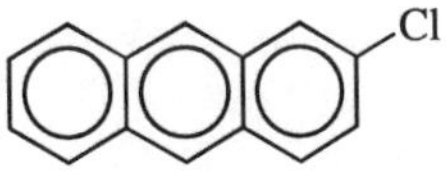 Cl	1.80	reversible at $v \leq 2000$ V/s
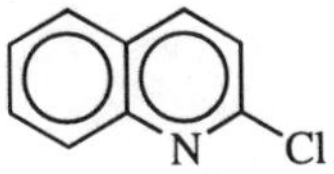 N, Cl	1.89	0.9
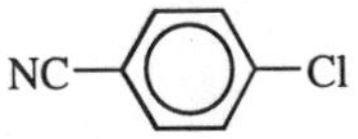 NC–, –Cl	2.08	0.6
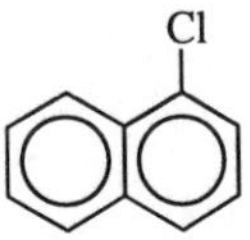 Cl	2.26	0.9

(continued)

Table 7. (continued)

Aryl bromides	$-E_i^0$ *(V versus SCE)*	"α" [b]
Cl, N	2.40	0.6
N, Cl	2.39	0.5
Cl	2.78	0.5
I	—	0.5
(*Aryl iodides*)		
Aryl Bromides		
O_2N, $CH(CH_3)_2$, Br	0.96	reversible at $v \leq 2000$ V/s
O_2N, CH_3, Br	0.98	reversible at $v \leq 2000$ V/s
O_2N, Br	0.98	reversible at $v \leq 2000$ V/s
O_2N, Br, CH_3	1.00	reversible at $v \leq 2000$ V/s
Br, NO_2	1.03	reversible at $v \leq 2000$ V/s
Br, O	1.19	reversible at $v \leq 2000$ V/s

(continued)

Table 7. (continued)

Aryl bromides	$-E_i^0$ *(V versus SCE)*	"α" [b]
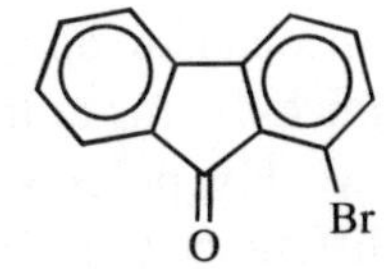 O, Br	1.20	reversible at $v \leq 2000$ V/s
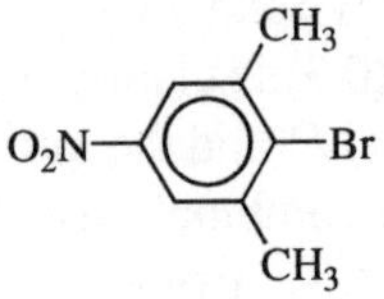 CH_3, O_2N–, –Br, CH_3	1.27	reversible at $v \leq 2000$ V/s
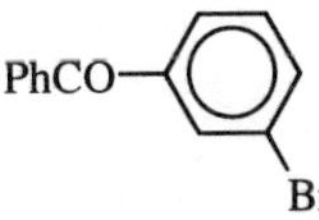 PhCO–, Br	1.53	reversible at $v \leq 2000$ V/s
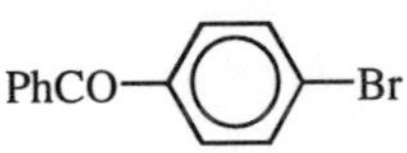 PhCO–, –Br	1.63	reversible at $v \leq 2000$ V/s
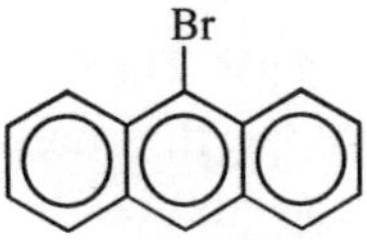 Br	1.70	0.9
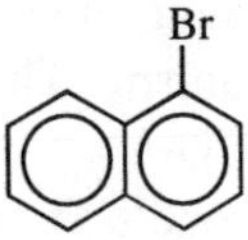 Br	2.19	0.6
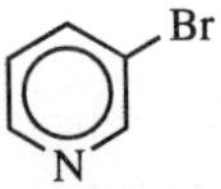 Br, N	2.29	0.5
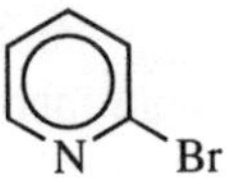 N, Br	2.30	0.5
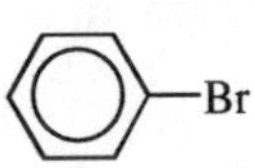 –Br	2.44	0.5

Notes: [a]In DMF + 0.1 M n-Bu_4NClO_4, at a mercury electrode at 22 °C.
[b]Average of the values derived from $\partial E_p / \partial \log v$ and $E_{p/2} - E_p$ between 0.1 and 100 V/s.

tials in the series and thus that α should not be less than 0.5 if the stepwise mechanism were followed.

The electrochemical reduction of trifluoromethyl bromide and iodide in DMF[27] is an interesting case in the sense that the mechanism appears to be of the concerted type while a $CF_3Br^{\bullet-}$ anion radical seems to have been detected upon γ-ray irradiation at 77 K in weakly polar solid matrixes (see below for a general discussion of the difference in mechanism that may result from the differences in the environmental factors and in the mode of the reduction). The value of α (0.3) is clearly below 0.5 with both CF_3Br and CF_3I. The standard potential E^0 and the intrinsic barrier $\Delta G_0^{\neq}$ of the reaction may be estimated from a combination of Eqs. (15), (2), and (26), leading to the determination of $\Delta G_0^{\neq}$ from

$$\Delta G_0^{\neq} = \frac{\Delta G_0^{\neq}}{4\alpha^2}$$

and E^0 from

$$E^0 + \phi_r = E_p + 4(1 - 2\alpha)\Delta G_0^{\neq}$$

It is found that $E^0 = -0.57$ V versus SCE and $\Delta G_0^{\neq} = 0.94$ eV. These values are not compatible with the intermediacy of $CF_3Br^{\bullet-}$ (perfluoroalkanes are reduced around –3V versus SCE and 0.94 eV appears too large to correspond merely to changes in bond lengths and reorganization of solvent) while they are with the dissociative mechanism. The same observations and conclusions were found to apply to CH_3I.

Similar observations and analyses lead one to conclude that the dissociative mechanism is also followed were made in an extended series of vicinal dibromides.

Most of the experimental studies in the field have concerned the cleavage of carbon–halogen bonds. Two exceptions worth noting are nitrogen–halogen bonds in aromatic *N*-halosultams and carbon–sulfur bonds in benzyl, benzhydryl, and triphenylmethyl phenyl (and *p*-nitrophenyl) sulfides.

Recent results indicate that the reductive cleavage of aromatic *N*-halosultams[28]

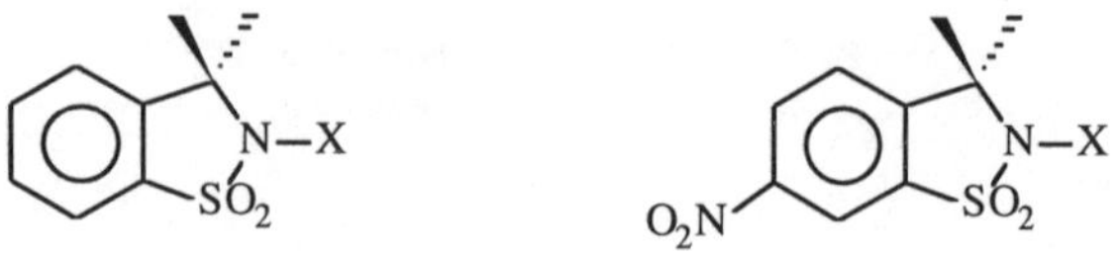

(where X = Br, Cl, I)

$$\rangle N{-}X + e^- \longrightarrow \rangle N^{\bullet} + X^-$$

follows the concerted mechanism in all cases but one, namely that of the nitrofluorine compound. In the latter case, $-\partial E_p/\partial \log v = 40$ mV, corresponding to an "α" of 0.75, whereas in all the other cases α is distinctly smaller than 0.5 and the location of the peak potential is not compatible with a stepwise mechanism.

In the arylmethylsulfides series,[29]

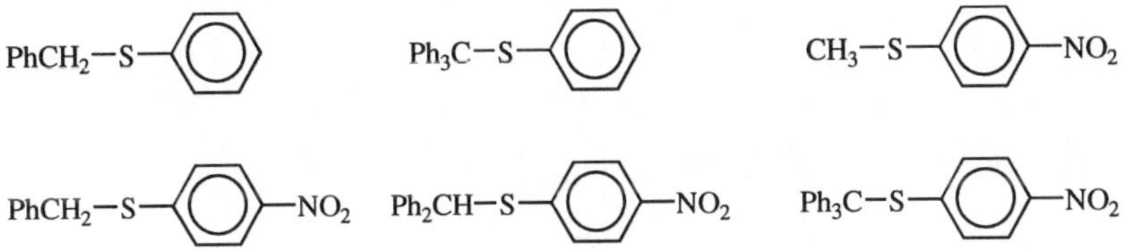

$PhCH_2{-}S{-}C_6H_5$ $Ph_3C{-}S{-}C_6H_5$ $CH_3{-}S{-}C_6H_4{-}NO_2$

$PhCH_2{-}S{-}C_6H_4{-}NO_2$ $Ph_2CH{-}S{-}C_6H_4{-}NO_2$ $Ph_3C{-}S{-}C_6H_4{-}NO_2$

the anion radicals of the *para*-nitroderivatives are so stable, with the exception of the triphenylmethyl derivative, that they can be observed by electrochemical *in situ* generation in ESR spectroscopy indicating lifetimes on the order of at least minutes. The anion radical of the triphenylmethyl *para*-nitrophenyl sulfide is much less stable, but the fact that one deals with a stepwise mechanism derives unambiguously from the observation that the cathodic peak potential varies by 30 mV per decade of the scan rate ("α" = 1). In the unsubstituted series, the electrochemical reduction is controlled by electron transfer. Although no unequivocal evidence for one or the other mechanism derives from the characteristics of the cyclic voltammogram, homogeneous electron transfer experiments point to a stepwise mechanism.

3.3. Discrimination Criteria in Homogeneous Reactions

As was stated before, a convenient way of investigating reductive cleavages by outer-sphere electron donors is to generate them electrochemically. The experiment starts with the introduction in the cyclic voltammetric cell of the oxidized form P of a reversible redox couple

P/Q, the standard potential of which, E^0_{PQ} , is positive to the reduction potential of RX. The loss of reversibility of the voltammogram and the increase of the cathodic peak height may then be used to extract quantitatively kinetic information[1] on the reaction of the homogeneous outer-sphere electron donor Q with RX :

$$P + e^- \rightleftharpoons Q$$

followed by

$$Q + RX \underset{k_{-i}}{\overset{k_i}{\rightleftharpoons}} P + RX^{\bullet -}$$

and

$$RX \xrightarrow{k_c} R^{\bullet} + X^-$$

in the case of a stepwise mechanism, or by

$$Q + RX \xrightarrow{k} P + R^{\bullet} + X^-$$

in the case of a concerted mechanism.

In the stepwise mechanism, there is a competition between the reverse electron transfer and the cleavage reaction; when $k_c << k_{-i}\, C^0_P$ (C^0_P is the bulk concentration of P) acts as a pre-equilibrium for the rate-determining step and the value of $(k_i/k_{-i})k_c$ can be extracted from the cyclic voltammetric measurements; when $k_c >> k_i C^0_P$, the rate-determining step is the forward electron transfer and k_i can be derived from the experimental data. The experimental criterion for distinguishing the first case from the second is the variation of the cyclic voltammetric response with the concentration C^0_P , given the value of the excess factor C^0_{RX} / C^0_P (C^0_{RX} is the bulk concentration of RX). In the first case the dimensionless kinetic parameter that governs the response, $(RT/Fv)\,(k_i k_c/k_{-i})$, where v is the scan rate does not depend on C^0_P, whereas in the second, the governing parameter $(RT/Fv)\, k_i\, C^0_P$ is proportional to C^0_P . The latter kinetic behavior is obtained with the dissociative mechanism.

A lack of variation of the electrochemical response from the concentration C^0_P therefore indicates that the stepwise mechanism is followed. Situations in which the parameter $k_{-i} C^0_P / k_c$ is neither large nor small not only allow one to infer the existence of the stepwise mechanism but also to determine the value of k_{-i}. Since k_{-i} is very often at the diffusion limit $RX^{\bullet-}$, lifetimes in the nanosecond range can be reached in this manner.[1,16]

In the case where the kinetic parameter is proportional to C_P^0 (i.e., when the electron transfer is rate-determining), there are still ways to decide whether the stepwise or the concerted mechanism is operating. The rate-determining electron transfer may then be governed by activation and/or diffusion:

$$\begin{array}{c} RX + Q \underset{k^{dif}}{\overset{k^{dif}}{\rightleftharpoons}} (RX + Q) \underset{k_{-i}^{act}}{\overset{k_i^{act}}{\rightleftharpoons}} (RX^{\bullet-} + P) \underset{k^{dif}}{\overset{k^{dif}}{\rightleftharpoons}} RX^{\bullet-} + P \\ (RX^{\bullet-} + P) \xrightarrow{k_c} R^{\bullet} + X^- + P \xleftarrow{k_c} RX^{\bullet-} + P \end{array} \tag{42}$$

where the parentheses represent the solvent cage, k_i^{act} and k_{-i}^{act} are the forward and reverse activation-controlled electron transfer rate constants, and k^{dif} is the bimolecular diffusion limit that can be approximated by the Smoluchovski expression

$$k^{dif} = 4\pi N_A(D_{RX} + D_Q)\,(a_{RX} + a_Q) \tag{43}$$

(the D's are the diffusion coefficients and the a's the hard-sphere radii). The dissociation reaction may occur outside the approximately spherical molecular diffusion layer as represented on the right-hand side of the above scheme or start inside the reaction layer. In the general case the following equation[30] applies :

$$\frac{1}{k_i} = \frac{1}{k_i^{act}} + \frac{1}{k^{dif}} + \frac{1}{k^{dif}\left[1 + \dfrac{a_{RX} + a_Q}{(D_{RX}+D_Q)^{1/2}}\,k_c^{1/2}\right]\exp\left[-\dfrac{F}{RT}\,(E_{PQ}^0 - E_i^0)\right]} \tag{44}$$

Figure 6 shows typical examples of the variation of the electron transfer rate constant k_i with the driving force as measured by $E_{PQ}^0 - E_i^0$ for a series of values of the intrinsic barrier $\Delta G_{0,i}^{\neq}$, taking account of the variation of k_i^{act} with the driving force :

$$k_i^{act} = Z^{hom}\exp\left[-\frac{F\Delta G_{0,i}^{\neq}}{RT}\left(1 + \frac{E_{PQ}^0 - E_i^0}{4\Delta G_0^{\neq}}\right)^2\right] \tag{45}$$

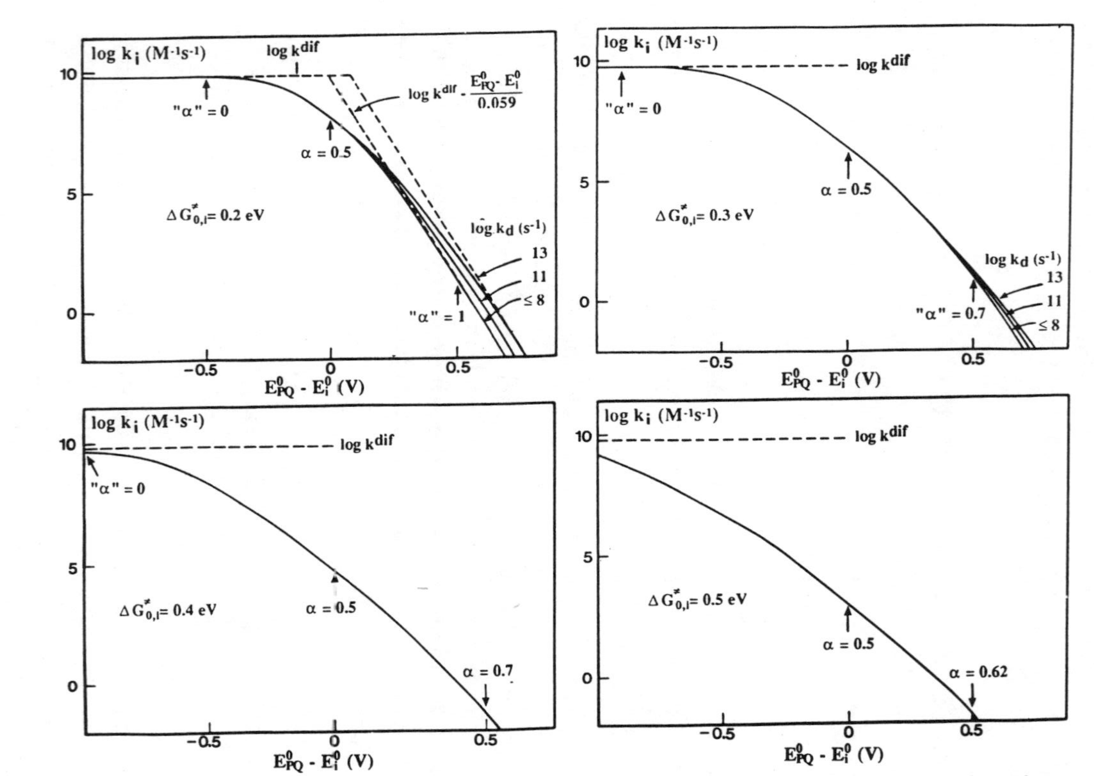

Figure 6. Stepwise mechanism in homogeneous reductive cleavages. Variation of the electron transfer rate constant k_i with the driving force $E^0_{PQ} - E^0_i$ for several values of the intrinsic barrier $\Delta G^{\neq}_{0,i}$ and of the follow-up dissociation rate constant k_c. Temp. = 25 °C, Z^{hom} = 3 × 10^{11} M^{-1} s^{-1}, k^{dif} = 6 × 10^9 M^{-1} s^{-1}, $a_{RX} + a_Q$ = 4Å,

In the accessible range of k_i values, it is seen that the slopes of $\log k_i - (E^0_{PQ} - E^0_i)$ that can be expressed as true or apparent symmetry factors, α or "α", are larger than 0.5.

On the other hand, with the concerted mechanism, α is smaller than 0.5, as illustrated by several theoretical $\log k - (E^0_{PQ} - E^0)$ curves in Figure 2.

Besides the criterion based on the value of the slope of the $\log k - E^0_{PQ}$ plots, one may also examine, as in the electrochemical case, whether or not the observed values of the rate constants are compatible with the stepwise mechanism based on estimated bracketing values of the standard potential E^0_i and the intrinsic barrier $\Delta G^{\neq}_{0,i}$.

3.4. Homogeneous Examples

The observation derived from direct electrochemistry that aromatic halides undergo a stepwise reductive cleavage is confirmed by experiments in which they are reacted with electrochemically generated aromatic anion radicals. Complete or partial kinetic control by the follow-up dissociation steps has, for example, been observed with 2-chloroquinoline (in DMF), 9-chloroanthracene, 9-bromoanthracene, 1-chloronaphthalene, 1-bromonaphthalene (in Me_2SO), and 4-chlorobenzonitrile (in MeCN)[26b] with chloro- and bromobenzene and 2- and 3-chloro- and bromopyridines. The kinetic control is by electron transfer but there is no ambiguity that this is the outer-sphere step and not the concerted cleavage as results from the possibility of fitting the experimental $\log k - E^0_{PQ}$ data by Eqs. (44) and (45) with corresponding "α" values between 0.5 and 1.[26a]

The same is obviously true for the vinyl halides investigated in Ref. 31, where the values of "α" much larger than 0.5. Figure 1 in Ref. 31 offers a particularly striking example of the difference between the behaviors expected for stepwise and concerted mechanisms: with 1.1-dichloroethane the value of α is 0.37 on the average, whereas with the vinyl halides it is only slightly below 1.

With the aliphatic and benzyl halides represented in Figure 3 we again find α values that are distinctly too low to correspond to the stepwise mechanism. The concerted mechanism (or even an S_N2 mechanism in the absence of steric hindrance) observed in direct electrochemistry is thus also followed in the homogeneous reduction by aromatic anion radicals.

3.5. Factors Governing the Passage from the Concerted to the Stepwise Mechanism

The experimental examples described earlier have shown that, even in the same family of compounds, the reductive cleavage may follow one or the other mechanism. The question thus arises of the nature of the factors that control the passage from one mechanism to the other.

At the level of driving forces, the occurrence of one or the other mechanism is governed by the potential difference

$$E^0 - E_i^0 = E^0_{RX/R^{\bullet}+X^-} - E^0_{RX/RX^{\bullet-}} \tag{46}$$

i.e.,

$$E^0 - E_i^0 = -D - T(\overline{S}_{RX} - \overline{S}_{R^{\bullet}} - \overline{S}_{X^{\bullet}}) + E^0_{X^{\bullet}/X^-} - E^0_{RX/RX^{\bullet-}} \tag{47}$$

(where the $\overline{S}$ are the partial molar entropies of the subscript species) : the more positive $E^0 - E_i^0$, the larger the preference for the concerted mechanism on thermodynamical grounds. It appears from the second expression for $E^0 - E_i^0$ that, given the departing group, the preference for the concerted mechanism is enhanced by small values of the R–X bond dissociation energy and by more and more negative values of the standard potential for the formation of the anion radical $E^0_{RX/RX^{\bullet-}}$. This $E^0_{RX/RX^{\bullet-}}$ is a measure of the free energy of the unpaired electron residing temporarily in an accessible unoccupied orbital of the RX molecule (for example a π^* orbital) before being transferred to the σ^* R – X orbital concertedly with the cleavage of the R – X bond.

One has, however, to take account of the fact that different activation barriers are to be overcome for generating the anion radical on one hand and for triggering the concerted reductive cleavage on the other. In general, the intrinsic barrier for the dissociative electron transfer is larger than that for the outer-sphere formation of $RX^{\bullet-}$, resulting in a kinetic disadvantage for the first reaction compared with the second for comparable driving forces. As discussed in detail elsewhere,[32] the dynamics of the anion radical may be modelled as follows: The potential energy curves involved in the stepwise and concerted mechanisms are schematically represented in Figure 7 as a function of the length z of the bond being broken. Moreover, z_R and z_i are the equilibrium values of z in the starting reactant (RX) and in the intermediate anion radical ($RX^{\bullet-}$), respectively. Based on the same assumptions as in the dissociative electron transfer model (RX and $RX^{\bullet-}$ are represented by Morse curves having the same

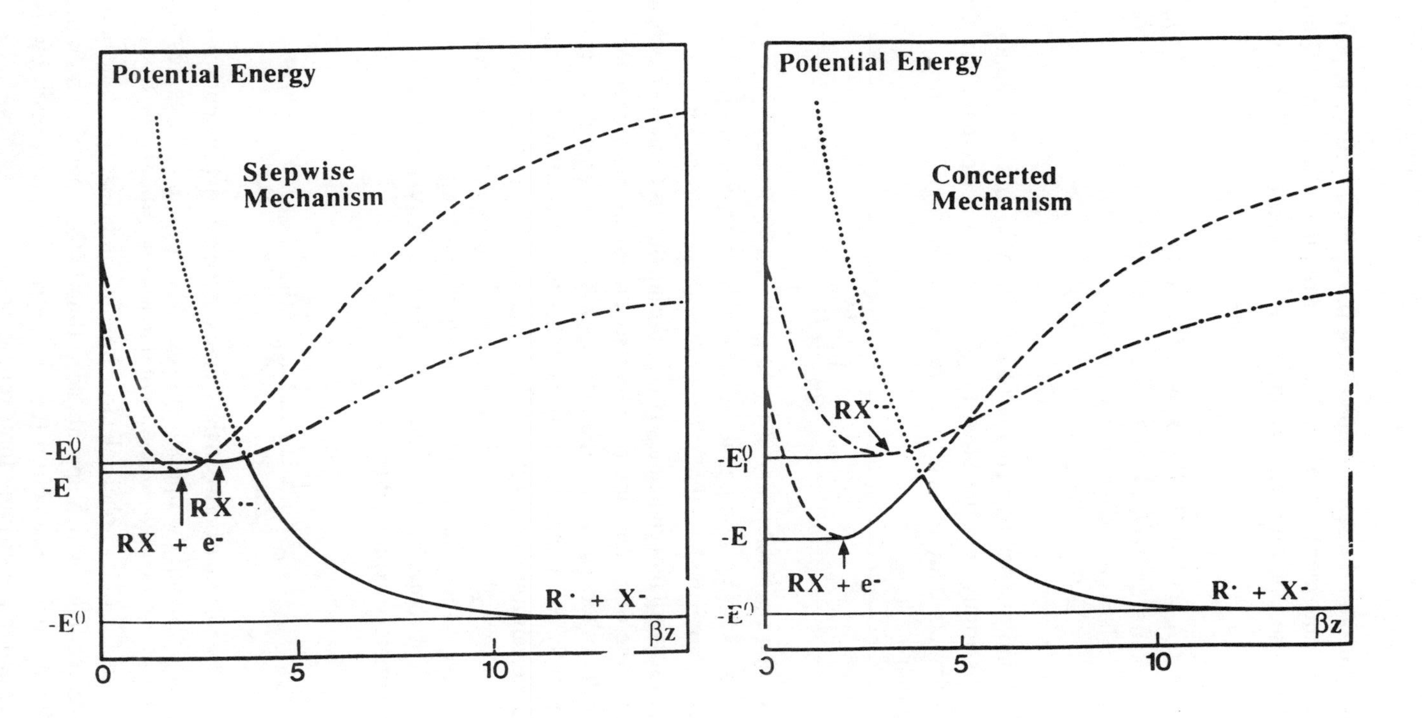

Figure 7. Passage from the stepwise to the concerted mechanism. Potential energy curves of RX [- - - , Eq. (48)], $RX^{\bullet -}$ [- · - , Eq. (49)], and $R^{\bullet} + X^{-}$ [. . . , Eq. (50)]. Solid line: reaction pathway.

repulsive part and the same shape factor β, the $R^\bullet$, X^- curve is purely dissociative and identical to the common repulsive part of RX and $RX^{\bullet-}$), and the potential energy curves are given by the following equations:

$$\begin{aligned} &RX && : G_R = G_R^0 + D\{1 - \exp[-\beta(z - z_R)]\}^2 && (48)\\ &RX^{\bullet-} && : G_i = G_i^0 + D_c\{1 - \exp[-\beta(z - z_i)]\}^2 && (49)\\ &R^\bullet + X^- && : G_P = G_P^0 + D\{\exp[-\beta(z - z_R)]\}^2 &&\\ & && \quad = G_P^0 + D_c\{\exp[-\beta(z - z_i)]\}^2 && (50) \end{aligned}$$

D_c is the bond dissociative energy of $RX^{\bullet-}$ into R^- and $X^\bullet$. The above assumptions imply that the bond lengths in the intermediate and the starting compound are related to the respective bond dissociation energies through

$$z_i - z_R = \frac{1}{2\beta}\ln\left(\frac{D}{D_c}\right) \tag{51}$$

indicating that the weakening of the bond in the anion radical as compared to the starting compound corresponds to an equilibrium bond length in the intermediate that is larger than in the starting reactant.

Most of solvent reorganization occurs during the dissociative electron transfer on the one hand and the outer-sphere electron transfer $RX \rightarrow RX^{\bullet-}$ on the other. However, the corresponding reorganization factors are not exactly the same. It is expected that the solvent reorganization factor λ_0 pertaining to the dissociative electron transfer be larger than that of the outer-sphere $RX \rightarrow RX^{\bullet-}$ reaction $\lambda_{0,i}$ and therefore that there is some solvent reorganization ($\lambda_{0,d}$) during the cleavage of $RX^{\bullet-}$.

The solvent reorganization factor $\lambda_{0,c}$ for the cleavage reaction may be approximately expressed as[32]

$$\lambda_{0,c} = e_0^2\left(\frac{1}{2a_{R^\bullet,X^-}} + \frac{1}{2a_{RX^{\bullet-}}} - \frac{1}{d}\right)\left(\frac{1}{D_{op}} - \frac{1}{D_s}\right) \tag{52}$$

(where d is the distance between the centers of the two equivalent spheres and D_{op} and D_s being the optical and static dielectric constants, respectively) . It follows that the dynamics of the three reactions—dissociative electron transfer, outer-sphere electron transfer, and cleavage of the bond in the intermediate (which appears according to this model as an intramolecular dissociative electron transfer)—may be represented by the following activation–driving force relationships:

$RX + e^- \rightarrow R^\bullet + X^-$,

$$\Delta G^{\neq} = w_R + \frac{D+\lambda_0}{4}\left(1+\frac{E-E^0-w_R+w_P}{D+\lambda_0}\right)^2 \quad (53)$$

For $RX + e^- \rightarrow RX^{\bullet -}$,

$$\Delta G_i^{\neq} = w_R + \frac{(D^{\frac{1}{2}} - D_c^{\frac{1}{2}})^2 + \lambda_{0,i}}{4}\left(1+\frac{E-E_i^0-w_R}{D^{1/2}-D_c^{1/2})^2+\lambda_{0,i}}\right)^2 \quad (54)$$

And for $RX^{\bullet -} \rightarrow R^\bullet + X^-$,

$$\Delta G_d^{\neq} = \frac{D_c+\lambda_{0,c}}{4}\left(1+\frac{E_i^0-E^0+w_P}{D_c+\lambda_{0,c}}\right)^2 \quad (55)$$

where the work terms w_R and w_P have the same definitions as in Section 2.1. In the first two equations, E represents either the electrode potential in the electrochemical case or the standard potential of the electron donor (previously denoted E^0_{PQ}) in the homogeneous case.

The dissociation energy of $RX^{\bullet -}$, D_c, is related to that of the starting compound D through

$$D_c = D - E^0_{R^\bullet/R^-} + E^0_{RX/RX^{\bullet -}} + T\,[(\bar{S}_{RX} - \bar{S}_{RX^{\bullet -}}) - (\bar{S}_{R^\bullet} - \bar{S}_{R^-})] \quad (56)$$

Neglecting the small entropic term, the weakening of the bond in $RX^{\bullet -}$ compared to RX amounts to the standard potential of $R^\bullet/R^-$ couple being positive with respect to that of the $RX/RX^{\bullet -}$ couple. The more so, the weaker the bond in $RX^{\bullet -}$ as compared to RX.

The intrinsic barrier of the dissociative electron transfer reaction is essentially governed by the RX bond dissociative electron transfer from both the thermodynamic and kinetic viewpoints. Changing D also modifies the intrinsic barrier of the $RX + e^- \rightleftharpoons RX^{\bullet -}$ electron transfer through the $(D^{1/2} - D_c^{1/2}D_c 1/2)^2$ term. Since $D > D_c$, a decrease in D results in an increase of the $RX + e^- \rightleftharpoons RX^{\bullet -}$ intrinsic barrier. As long as the cleavage reaction is not so fast as to make the $RX + e^- \rightarrow RX^{\bullet -}$ the rate-determining step of the stepwise pathway, the above effect is opposed by the acceleration of the cleavage reaction resulting from a decrease in D (its driving force increases, and its intrinsic barrier decreases). Because this compensating effect tends to level off as D decreases, it does not reverse the thermodynamic advantage of the dissociative electron transfer caused by a decrease of D. More precise estimations of the intrinsic barrier of the cleavage reaction may be found in Ref. 32.

The other main controlling factor is the value of $E^0_{RX/RX^{\bullet-}}$. A more and more negative value of $E^0_{RX/RX^{\bullet-}}$, not only favors the dissociative electron transfer thermodynamically, but, since it decreases D_c, tends to slow down the $RX + e^- \rightleftharpoons RX^{\bullet-}$ reactions. As before, this effect may be compensated for up to a point by the ensuing acceleration of the reductive cleavage, again not reversing the thermodynamic trend.

These observations allow one to reconcile the various experimental findings reported in Sections 3.2 and 3.4.

The fact that the reductive cleavage of all investigated aryl halides follows the stepwise mechanism while benzylic halides do not is essentially a reflection of a larger bond strength: for comparable values of $E^0_{RX/RX^{\bullet-}}$, the dissociative electron transfer has a driving force advantage of ca. 1 eV[33] in the benzylic case as compared to the aryl case. It is noteworthy that for most of the aryl halides listed in Table 7, the dissociative electron transfer has no driving force advantage over the stepwise reaction (i.e., the anion radical is thermodynamically stable).

As compared to aryl halides, simple aliphatic halides have both a much more negative $E^0_{RX/RX^{\bullet-}}$ and a weaker bond energy.

In the benzylic series, the fact that the nitro-substituted derivatives undergo a stepwise reductive cleavage whereas all the other compounds follow a concerted mechanism derives essentially from the more positive value of $E^0_{RX/RX^{\bullet-}}$ in the former case than in the latter. The dissociative electron transfer driving force advantage resulting from a decrease in $E^0_{RX/RX^{\bullet-}}$ is compensated only weakly by the opposing increase in D.[12]

A similar trend is found in the aromatic N-halosultam series (see Section 3.2)[28] when comparing the nitro-substituted N-F derivative (stepwise mechanism) to the unsubstituted N-F derivative (concerted mechanism) in accordance with the fact that $E^0_{RX/RX^{\bullet-}}$ is ca. 1.4 V more positive in the first case than in the second. In the same family of compounds, another example illustrates the role of the bond dissociation energy: in the nitro-substituted series the N–Cl derivative undergoes a concerted reductive cleavage, whereas the N–F derivative follows the stepwise mechanism in accordance with the fact that the bond dissociative energy is larger in the first case than in the second.

In most cases, the conclusions of direct or indirect electrochemical studies such as those described above agree with the conclusions derived from gas-phase studies or γ-ray irradiation in apolar or weakly polar solid matrixes at low-temperature (usually 77 K), although in the latter case it is sometimes difficult to distinguish between a weakly bound $R^{\bullet},X^-$ adduct and a true $RX^{\bullet-}$ anion radical. With these techniques, anion radicals are found with aryl halides[34] and not with aliphatic halides.[34c,35] Pulse radiolysis studies also point to the formation of the anion radical with the aryl halides in water.[36]

An interesting borderline case is that of perfluoroalkyl halides, where the gas-phase and low-temperature matrix studies indicate the intermediacy of the anion radical,[37] whereas the electrochemical reduction appears to follow the concerted mechanism as discussed in Section 3.2. As shown by recent theoretical calculations,[4c] this difference in behavior is related to the polar character of the solvent used in the electrochemical studies.

In the case of nitrobenzyl halides, anion radicals are found as intermediates on γ-ray irradiation in solid matrixes at low temperature[38] as well as in pulse radiolysis in water.[39] There is, however, a striking difference between the electrochemical reductive cleavage of the 3-cyanobenzyl bromide in DMF and its reductive cleavage by pulse radiolysis in water.[39c] The anion radical is observed in pulse radiolysis in water[39c] with a decay-rate constant of 1.3×10^7 s^{-1} whereas, as discussed before, electron transfer and bond breaking are concerted during the electrochemical reduction in DMF. One reason for this difference in behavior may be that the driving force of the reaction is much larger in the first case than in the second: the standard potential of the solvated electron in water has been estimated as ca. –3.11 V versus SCE,[40] whereas the cyclic voltammetric reduction takes place under much milder conditions, since the peak potential varies between –1.53 V versus SCE at 0.1 V s^{-1} and –1.74 V versus SCE at 10 V s^{-1}. One additional reason may be that the anion radical is likely to be less kinetically stable in DMF than in water. In this connection we note that, at room temperature, the lifetime of the anion radical of 4-nitrobenzyl chloride in water (250 μs) is shorter by three orders of magnitude than that in DMF (0.25 μs). A possible explanation of this large difference is as follows: In the anion radical a considerable portion of the negative charge is located on the oxygen atoms of the nitro group. A strong interaction of this portion of the anion radical with water molecules is thus likely. This is expected to have two main effects on the kinetics of the cleavage reaction. One is to lower the energy of the π^* orbital in which the unpaired electron is mostly located and therefore to decrease the driving force of the cleavage reaction, as discussed earlier. The other effect has to do with solvent reorganization, which ought to be large since the solvation of the NO_2^- group by water molecules has to be changed into a solvation of the departing chloride ion. It is remarkable in this connection that very large pre-exponential factors have been found in the cleavage kinetics of nitro-substituted benzyl halide anion radicals in water,[39d] implying a positive entropy of activation.

As appears in Figure 7, with borderline systems, a simple change of the driving force may result in a change of mechanism in the reductive cleavage of a given molecule: a decrease in the driving force then makes the mechanism pass from stepwise to concerted. There have been few experimental investigations of this aspect of the stepwise-versus-concerted problem.

The reductive cleavage of triphenylmethylphenyl sulfide offers an interesting attempt in this connection.[29d] The electrochemical reduction is charge-transfer controlled with a transfer coefficient smaller than 0.5 (0.4) consistent with a concerted pathway. In the reduction by aromatic anion radicals, the variations of the rate constant with the driving force, as represented by the standard potential of the aromatic anion radicals (see Figure 3 in Ref. 29d), show a small but distinct change of the curve on which the data points are located following a decrease in the driving force of the reaction, as predicted for the passage from a stepwise to a concerted pathway: after the standard potential of the electron donor is raised, the reaction is first controlled by the kinetics of the $RX + e^- \rightarrow RX^{\bullet -}$ step, then by the follow-up cleavage $RX^{\bullet -} \rightarrow R^{\bullet} + X^-$, and finally by the dissociative electron transfer $RX + e^- \rightarrow R^{\bullet} + X^-$.

Such a change of mechanism, triggered by a variation of the energy of the incoming electron, has recently been found in the reduction of dialkyl aryl sulfonium cations[41] and of vicinal dibromide.[42]

4. CONCLUSIONS

Criteria for distinguishing concerted from stepwise mechanisms in electrochemical and homogeneous reductive cleavage reactions in polar solvents have been successfully applied in a number of cases. The main factors that govern the occurrence of one or the other mechanism appear to be the bond dissociation energy, the standard potential of the $RX/RX^{\bullet -}$ (i.e., the energy of the orbital in which the unpaired electron is liable to reside transitorily), and the free energy of the electron donor. Additional experimental examples showing the effect of the latter parameter on the transition between the two mechanisms in the reductive cleavage of a given molecule would certainly be welcome. Besides the structure of the starting compound, the nature of the reaction medium (solvent, ion pairing with countercations) most probably plays an important role in the determination of the mechanism, mostly through the ensuing variations of the $RX/RX^{\bullet-}$ standard potential.

When the concerted mechanism prevails, the general and approximate Morse curve model has been shown to describe with reasonable accuracy the dynamics of the reaction for several organic molecules. It may thus be used to estimate bond dissociation energies, standard potentials, and intrinsic barriers for dissociative electron transfer reactions involving one-electron homogeneous or heterogeneous electron donors and cleaving acceptors for which the necessary thermochemical data are presently lacking. The experimental examples investigated so far have mostly involved carbon and nitrogen–halogen bonds. Extension to other molecules, including small inorganic molecules, is certainly warranted.

Among the improvements in the theory that the multiplication of experimental examples will most probably call for, some may already be envisioned. For example, the assumption that the contributions of bond breaking and of solvent reorganization may be treated separately and additively is rather crude and may appear as a significant shortcoming in cases where the latter factor plays a relatively more important role than in the cases investigated so far, where the contribution of solvent reorganization is not more than ca. 20% of the total. Refinement of the theory, including the modelling of the dependency of the solvent reorganization on the bond length, may then appear necessary. Another problem, which may at first sight appear to concern solely electrochemical cleavages, is the influence of the intense electric field that exists at the reaction site on the dynamics of the dissociative electron transfer. Similar questions may arise in homogeneous reactions in the form of salt effects.

ACKNOWLEDGMENTS

I am indebted to all the co-workers who participated to our own work on dissociative electron transfer and whose names may be found in the reference list. The invaluable contribution of Dr. Claude P. Andrieux to this field should be particularly emphasized.

REFERENCES

1. Andrieux, C. P.; Savéant, J-M. In *Electrochemical Reactions in Investigation of Rates and Mechanisms of Reactions, Techniques of Chemistry*; Bernasconi, C. F., Ed.; Wiley: New York, 1986; Vol. VI/4E, Part 2, pp 305–390.

2. (a) Beugelmans, R. *Bull. Soc. Chim. Belg.* **1984**, *93*, 547. (b) Bowman, W. R. In *Photoinduced Electron Transfer*; Fox, M. A. and Channon, M., Eds.; Elsevier: Amsterdam, 1988; Part C, pp 487–552. (c) Bowman, W. R. *Chem. Soc. Rev.* **1988**, *17*, 283. (d) Bunnett, J. F. *Acc. Chem. Res.* **1978**, *11*, 413. (e) Kornblum, N. *Angew. Chem.* **1975**, *14*, 734. (f) Kornblum, N. In *The Chemistry of Functional Groups*; Patai, S., Ed.; Wiley: New York, 1982; Supplement F., p 361. (h) Lablache-Combier, A. In *Photoinduced Electron Transfer*, Fox, M. A. and Channon, M., Eds.; Elsevier: Amsterdam, 1988; Part C, pp 134–312. (i) Norris, R. K. In *The Chemistry of Functional Groups*; Patai, S. and Rappoport, Z., Eds.; Wiley: New York, 1983; Supplement D'., Chapter 16, p 681. (k) Norris, R. K. In *Comprehensive Organic Chemistry*, Trost, M., Ed.; Pergamon: New York, 1992; Vol. 4, Chapter 2.2, pp 451–482. (l) Rossi, R. A.; Rossi, R. H. *Aromatic Substitution by the $S_{RN}1$ Mechanism: ACS Monograph 178*; The American Chemical Society: Washington D.C., 1983. (m) Rossi, R. A.; Pierini, A. B.; Palacios, S. M. *Adv. Free Rad. Chem.* **1990**, *1*, 193. (n) Russell, G. A. *Spec. Publ. Chem. Soc.* **1970**, *24*, 271. (o) Russell, G. A. In *Advances in Physical Organic Chemistry*, Bethel, D., Ed.; Academic Press: London, **1987**, *24*, 271. (q) Russell, G. A. *Acc. Chem Res.* **1989**, 22, 1. (r) Savéant, J-M. *Acc. Chem Res.* **1980**, *13*, 323. (s) Savéant, J-M. *Adv. Phys. Org. Chem.* **1990**, *26*, 1. (t) Wolfe, J. F.; Carver, D. R. *Org. Prep. Proc. Int.* **1978**, *10*, 225. (u) Savéant, J-M. *Tetrahedron Report,* in press.
3. (a) Marcus, R. A. *J. Chem. Phys.* **1956**, *24*, 4966. (b) Hush, N. S. *J. Chem. Phys.* **1958**, *28*, 962. (c) Marcus, R. A. In *Special Topics in Electrochemistry*; Rock, P. A., Ed.; Elsevier: New York, 1977; pp 161–179. (d) Marcus, R. A. *Faraday Discuss. Chem. Soc.* **1982**, *74,* 7. (e) Marcus, R. A.; Sutin, N. *Biophys. Biochim. Acta* **1985**, *811*, 265.
4. (a) Savéant, J-M. *J. Am. Chem. Soc.* **1987**, *109*, 6788. (b) Wentworth, W. E.; George, R.; Keith, H. *J. Chem. Phys.* **1969**, *51*, 1791. (c) Bertran, J.; Gallardo, I.; Moreno, M.; Savéant, J-M., *J. Am. Chem. Soc.* **1992**, *114*, 9576. (d) Savéant, J-M. *J. Am. Chem. Soc.* **1992**, *114*, 10595.
5. (a) Sutin, N. In *Progress in Inorganic Chemistry*; Lippard, S. J., Ed.; Wiley: New York, 1983, Vol. 30, pp 441–447. (b) Newton, M. D.; Sutin, N. *Ann. Rev. Phys. Chem.* **1984**, *35*, 437. (c) Sumi, H.; Marcus, R. A. *J. Chem. Phys.* **1986**, *84*, 4272. (d) Sumi, H.; Marcus, R. A. *J. Chem. Phys.* **1986**, *84*, 4894. (e) Sumi, H.; Marcus, R. A. *J. Electroanal. Chem.* **1986**, *204*, 59. (f) Nadler, W.; Marcus, R. A. *J. Chem. Phys.* **1987**, *86*, 3906. (g) Jortner, J.; Bixon, M. *J. Chem. Phys.* **1988**, *88,* 167. (h) Weaver, M.; McMannis, G. E. *Acc. Chem. Res.* **1990** *23*, 294. (i) Grampp, G.; Jänicke, W. *Ber. Bunsenges. Phys. Chem.* **1991**, *95*, 904. (j) Fawcett, W. R.; Blum, L. *Chem. Phys. Lett.* **1991**, *187*, 173.
6. Andrieux, C. P.; Gallardo, I.; Savéant, J-M.; Su, K. B. *J. Am. Chem. Soc.* **1986**, *108*, 638.
7. (a) Delahay, P. *Double Layer and Electrode Kinetics*, Wiley: New York, 1965. (b) Kojima, H.; Bard, A. *J. Am. Chem. Soc.* **1975**, *97*, 6317.
8. (a) Lexa, D.; Savéant, J-M.; Su, K. B.; Wang, D. L. *J. Am. Chem. Soc.* **1987**, *109*, 6464. (b) Andrieux, C. P.; Savéant, J-M. *J. Electroanal. Chem.* **1989**, *267*, 15. (c) Lexa, D.; Savéant, J-M.; Schäfer, H. J.; Su, K. B.; Vering, B.; Wang, D. L. *J. Am. Chem. Soc.* **1990**, *112*, 271.

9. (a) Hush, N. S. Z. *Electrochem.* **1957**, *61*, 734. (b) Eberson, L. *Acta Chem. Scand.* **1982**, *B36*, 533.
10. (a) Cox, B. G.; Hedvig, G. R.; Parker, A. J.; Watts, D. W. *Aust. J. Chem.* **1974**, *27*, 477. (b) Geske, D. H.; Ragle, M. A.; Bambeneck, J. L.; Balch, A. L. *J. Am. Chem. Soc.* **1964**, *86*, 987. (c) Benson, S. W. *Thermodynamical Kinetics*, 2nd ed.; Wiley: New York, 1976. (d) Wagman, D. D.; Evans, W. H.; Parker, V. B.; Schumm, R. H.; Halow, I.; Bailey, S. M.; Churney, K. L.; Nuttal, R. L. *J. Phys. Chem. Ref. Data.* **1982**, *11*, Supplement 2.
11. Savéant, J-M.; Tessier, D. *Faraday Discuss. Chem. Soc.* **1982**, *74*, 57.
12. Andrieux, C. P.; Le Gorande, A.; Savéant, J-M. *J. Am. Chem. Soc.* **1992**, *114*, 6892.
13. Clark, K. B.; Wayner, D. D. M. *J. Am. Chem. Soc.* **1991**, *113*, 9363.
14. Adcock, W.; Clark, C. I.; Houman, A.; Krstic, A. R.; Pinsen, J.; Savéant, J-M.; Taylor, D. K.; Taylor, J. F. *J. Am. Chem. Soc.* **1994**, *116*, 4653.
15. (a) Lexa, D.; Mispelter, J.; Savéant, J-M. *J. Am. Chem. Soc.* **1981**, *103*, 6806. (b) Lund, T.; Lund, H. *Acta Chem. Scand.* **1986**, *B40*, 470. (c) Lexa, D.; Savéant, J-M.; Su, K. B.; Wang, D. L. *J. Am. Chem. Soc.* **1988**, *110*, 7617. (d) Daasbjerg, K.; Pedersen, S. U.; Lund, H. *Acta Chem. Scand.* **1991**, *24*, 470.
16. Andrieux, C. P.; Hapiot, P.; Savéant, J-M. *Chem. Rev.* **1990**, *90*, 723.
17. Lund, T.; Lund, H. *Acta Chem. Scand.* **1987**, *B41*, 93.
18. Hebert, E.; Mazaleyrat, J. P.; Nadjo, L.; Savéant, J-M.; Welvart, Z. *Nouv. J. Chem.* **1985**, *9*, 75.
19. (a) Lund, T.; Lund, H. *Acta Chem. Scand.* **1987**, *B41*, 93. (b) Lund, T.; Lund, H. *Acta Chem. Scand.* **1988**, *B42*, 269.
20. (a) Bordwell, F. G.; Hughes, D. L. *J. Org. Chem.* **1983**, *48*, 2206–2215. (b) Bordwell, F. G.; Bausch, M. J.; Wilson, C. A. *J. Am. Chem. Soc.* **1987**, *109*, 5465–5470. (c) Bordwell, F. G.; Wilson, C. A. *J. Am. Chem. Soc.* **1987**, *109*, 5470–5474. (d) Bordwell, F. G.; Harrelson, J. A. Jr. *J. Am. Chem. Soc.* **1987**, *109*, 8112–8113. (e) Bordwell, F. G.; Harrelson, J. A. Jr. *J. Am. Chem. Soc.* **1989**, *111*, 1052–1057.
21. (a) Cho, J. K.; Shaik, S. S. *J. Am. Chem. Soc.* **1991**, *113*, 9890. (b) Eberson, L.; Shaik, S. S. *J. Am. Chem. Soc.* **1990**, *112*, 4484.
22. Peover, M. I.; Powell, J. S. *J. Electroanal. Chem.* **1969**, *20*, 427.
23. Morgan, J. D.; Wolynes, P. G. *J. Phys. Chem.* **1987**, *91*, 874.
24. Grimshaw, J.; Langan, J. R.; Salmon, G. A. *J. Chem. Soc. Chem. Commun.* **1988**, 1115.
25. Nadjo, L.; Savéant, J-M. *J. Electronanal. Chem.* **1973**, *48*, 113.
26. (a) Andrieux, C. P.; Blocman, C.; Dumas-Bouchiat, J-M.; Savéant, J-M. *J. Am. Chem. Soc.* **1979**, *101*, 3431. (b) Andrieux, C. P.; Blocman, C.; Dumas-Bouchiat, J-M.; M'Halla, F.; Savéant, J-M. *J. Am. Chem. Soc.* **1980**, *102*, 3806. (c) Andrieux, C. P.; Savéant, J-M.; Zann, D. *Nouv. J. Chim.* **1984**, *8*, 107.
27. Andrieux, C. P.; Gélis, L.; Medebielle, M.; Pinson, J.; Savéant, J-M. *J. Am. Chem. Soc.* **1990**, *112*, 3509.
28. Andrieux, C. P.; Differding, E.; Robert, M.; Savéant, J-M. *J. Am. Chem. Soc.* **1993**, *115*, 6592.
29. (a) Farnia, G.; Severin, M. G.; Capobianco, G.; Vianello, E. *J. Chem. Soc. Perkin Trans.* **1978**, *2*, 1. (b) Capabianco, G.; Farnia, G.; Severin, M. G.; Vianello, E. *J.*

Electroanal. Chem. **1984**, *165*, 251. (c) Severin, M. G.; Arévalo, M. C.; Farnia, G.; Vianello, E. *J. Phys. Chem.* **1987**, *91*, 466. (d) Severin, M. G.; Farnia, G.; Vianello, E.; Arévalo, M. C. *J. Electroanal. Chem.* **1988**, *251*, 369

30. (a) Andrieux, C. P.; Savéant, J-M. *J. Electroanal. Chem.* **1986**, *205*, 43. (b) Grimshaw, J.; Thompson, N. *J. Electroanal. Chem.* **1986**, *205*, 35.
31. Gatti, N.; Jugelt, W.; Lund, H. *Acta Chem. Scand.* **1987**, *B41*, 646.
32. Savéant, J-M. *J. Phys. Chem.* **1994**, *98*, 3716.
33. *Handbook of Chemistry and Physics*, 72nd Edition, CRC: Boca Raton, 1991–1992, pp. 9–121.
34. (a) Wentworth, W. E.; Becker, R. S.; Tung, R. *J. Phys. Chem.* **1967**, *71*, 1652. (b) Steelhammer, J. C.; Wentworth, W. E. *J. Chem. Phys.* **1969**, *51*, 1802. (c) Symons, M. C. R. *Pure Appl. Chem.* **1981**, *53*, 223.
35. Compton, R. N.; Reinhart, P. W.; Cooper, C. C. *J. Chem. Phys.* **1978**, *68*, 4360.
36. (a) Neta, P.; Behar, D. *J. Am. Chem. Soc.* **1981**, *103*, 103. (b) Neta, P.; Behar, D. *J. Am. Chem. Soc.* **1981**, *103*, 2280. (c) Bays, J. P., Blumer, S. T.; Baral-Tosh, S.; Behar, D.; Neta, P. *J. Am. Chem. Soc.* **1983**, *105*, 320.
37. (a) Hasegawa, A.; Williams, S. *Chem. Phys. Lett.* **1977**, *46*, 66. (b) Hasegawa, A.; Shiatani, M.; Williams, S. *Faraday Discuss. Chem. Soc.* **1977**, *157*. (c) Kuhn, A.; Illenberger, E. *J. Phys. Chem.* **1989**, *93*, 7060. (e) Kuhn, A.; Illenberger, E. *J. Chem. Phys.* **1990**, *93*, 357.
38. Symons, M. C. R.; Bowman, W. R. *J. Chem. Soc. Chem. Commun.* **1984**, 1445.
39. (a) Neta, P.; Behar, D. *J. Am. Chem. Soc.* **1980**, *102*, 4798. (b) Norris, R. K.; Barker, S. D.; Neta, P. *J. Am. Chem. Soc.* **1984**, *106*, 3140. (c) Meot-Ner, M.; Neta, P.; Norris, R. K.; Wilson, K. *J. Phys. Chem.* **1986**, *90*, 168.
40. Hart, E. J.; Anbar, M. *Hydrated Electron*, Wiley: New York, 1970.
41. Andrieux, C. P.; Robert, M.; Saeva, F. D.; Savéant, J-M. *J. Am. Chem. Soc.*, in press.
42. Andrieux, C. P.; de Goranda, A.; Savéant, J-M. *J. Electroanal. Chem.*, in press.

PHOTOINDUCED ELECTRON TRANSFER CHEMISTRY OF AMINES AND RELATED ELECTRON DONORS

Ung Chan Yoon, Patrick S. Mariano,
Richard S. Givens, and Beauford W. Atwater III

Advances in Electron Transfer Chemistry
Volume 4, pages 117–205.

ISBN: 1-55938-506-5

1. INTRODUCTION

1.1. General Considerations

Photoinduced electron transfer (SET) has become a topic of considerable interest, as demonstrated by the large number of recent reviews summarizing its basic concepts[1–8] and its mechanism and synthetic applications.[9–12] The recognition that excited-state reactions can be initiated by SET in donor–acceptor pairs has had a significant impact on organic photochemistry. It has led to new ways of analyzing and designing photochemical processes of organic substances and of developing novel strategies for chemical synthesis. This chapter focuses on excited-state chemical transformations of this type and, specifically, on photoinduced SET reactions in which amines and related substances serve as electron donors. Although our scope has been narrowed by the selection of this specific class of substrates, the topics chosen for coverage permit us to emphasize many of the major concepts involved in understanding the intricate mechanisms of excited-state SET chemistry and in predicting preparatively useful processes. Thus, our aim here is to summarize general principles rather than to present a thorough review of all of the interesting chemistry uncovered. Finally, a major portion of this chapter is devoted to our own studies in this area, which have been conducted during the past decade.

1.2. Concepts of Photochemical SET

Photoinduced SET processes (Scheme 1) are initiated by the transfer of an electron from or to excited-state species to or from respective ground-state acceptors or donors. As a consequence, the types of reaction pathways followed as well as their efficiencies are governed by the redox properties of the donor and acceptor partners, the energy of the excited states participating in the processes, and the rates and natures of secondary reactions of the charged or neutral radicals formed in both the SET and subsequent steps. Thus, SET-promoted photochemical reactions often contrast dramatically with classical (e.g., *cis–trans* isomerizations, rearrangements) excited processes in which the electronic properties (e.g., bond order, charge densities) of the excited states control the type of chemistry occurring.

It is clear from viewing the general sequence given in Scheme 1 that transformations of substrates M to products P_2 of SET–photochemical pathways depend on a number of critical factors. Firstly, SET routes (k_{SET}) are often competitive with classical decay (k_d) and reaction (k_{es}) modes available to the excited reactant M^*. Owing to this, partitioning of M^* in the direction of product (P_2) formation is determined by the SET rate constant k_{SET}. Both theoretical[13,14] and empirical[15] relationships are now available to estimate k_{SET} with reasonable accuracy based on the excitation energy of M^* and the ground-state redox properties of M and the acceptor A or donor D. Thus, for modestly exothermic (ΔG^o_{SET} = ca. –5 to –40 kcal/mol) bimolecular processes, SET is diffusion controlled (i.e., k_{SET} = ca. k_{diff} = ca. $1 \times 10^{10}\ M^{-1}\ s^{-1}$).

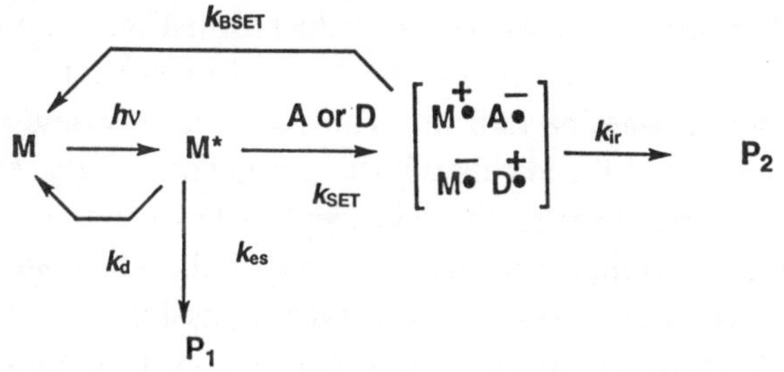

k_{BSET}
M $\xrightarrow{h\nu}$ M^* $\xrightarrow[k_{SET}]{\text{A or D}}$ $[M^{\bullet +}\ A^{\bullet -}]$ / $[M^{\bullet -}\ D^{\bullet +}]$ $\xrightarrow{k_{ir}}$ P_2
k_d
k_{es}
P_1

Scheme 1.

Secondly, the ion–radical pairs generated in the SET steps are highly reactive intermediates and, depending on their nature (contact versus solvent-separated), multiplicities (singlet versus triplet), and energies (i.e., ΔG_{SET} for electron backtransfer), electron backtransfer (k_{BSET}) can serve as an energy-wasting (-quenching) pathway. Consequently, the efficiencies of photoinduced SET processes leading to P_2 are a sensitive function of the rates k_{ir} of forward reactions available to these charged radical intermediates. Moreover, the chemical efficiencies (selectivities) of reactions proceeding by these sequences are controlled by the relative rates of competing ion–radical processes. Owing to their interesting and potentially well understood ion- and radical-like properties, the charged radical intermediates formed in photoinduced SET routes often react in patterned and predictable ways. As a result, one can apply simple organic-chemical reasoning to design high and selective reactivity into these intermediates.[2]

1.3. Overview of Coverage

As mentioned above, the intent of this chapter is to review photoinduced SET reactions in which amines and their derivatives serve as electron donors. Earlier discussions of this topic have been published.[16,17] Consequently, our coverage will begin with only a brief discussion of the redox properties of amines and, in this way, with an evaluation of the types of acceptors that are capable of participating in efficient SET processes with these substances. Since amine cation radicals (aminium radicals) serve as key reactive intermediates in the reaction pathways followed, relevant chemical properties of these short-lived species will be summarized. We will attempt to provide an understanding of the types of reactions these intermediates typically undergo and to evaluate the factors that influence the relative rates of potentially competitive pathways. Furthermore, in pathways for product formation, aminium radicals frequently serve as precursors of neutral, nitrogen-centered aminyl or carbon-centered α-amino radicals. Thus, a short overview of the properties of these neutral radicals will be provided. Finally, excited-state SET from amine donors to neutral acceptors leads to the generation of acceptor-derived anion radicals. In most of the chemistry described below, the acceptor anion radical serves a passive role and does not influence the nature of the aminium radical transformation occurring. Thus, the properties of these negatively charged odd-electron intermediates will not be reviewed separately, but only in the context of processes where they play an important role in determining the nature of the reaction pathways followed.

2. BACKGROUND

2.1. Redox Properties of Amines

SET photoreactions in which amines serve as ground-state electron donors [Eq. (1)] are initiated by redox processes in which the thermodynamic facilities (ΔG^{o}_{SET}) are governed by the amine oxidation potentials [$E_{\frac{1}{2}}(+)$] and the excited acceptor reduction potentials [$E_{\frac{1}{2}}^{*}(-) = E_{0,0}^{*} + E_{\frac{1}{2}}(-)$, where $E^{*}_{0,0}$ is the excited-state energy and $E_{\frac{1}{2}}(-)$ the reduction potential of the acceptors] in the following manner: $\Delta G_{SET}^{o} = E_{\frac{1}{2}}(+) - E_{\frac{1}{2}}^{*}(-)$. As anticipated, amines are among the most easily oxidized of neutral organic substances, as reflected by their low oxidation potentials (Table 1).[18] The observed effects of *N*-alkyl [i.e., $E_{\frac{1}{2}}(+)$ decrease in series 1°, 2°, 3°] and *N*-acyl [$E_{\frac{1}{2}}(+)$ increase in series R_3N, R_2NCO_2R, R_2NCOR] substitution are easy to forecast. Pertinent to the discussion below is the effect of α-substitution by metal containing ligands such as R_3Si and R_3Sn on amine $E_{\frac{1}{2}}(+)$ values. Yoshida[22] and Cooper[23] have demonstrated that α-silylamines and -carbamates are more easily oxidized than their non-silicon-containing counterparts (Table 1). This phenomenon can be understood in terms of the stabilizing effect of delocalization of odd-electron and positive charge density in the α-silylaminium radical into the carbon–silicon bond as a consequence of the energetically favorable N_p and C–Si_{σ} orbital interaction. Trends such as these have been noted and explained in a similar fashion for allyl- and benzyl silanes[24,25] and thioethers.[26]

$$\text{Acceptor} \xrightarrow{h\nu} \text{Acceptor}^{*} \underset{R_3N}{\overset{SET}{\longrightarrow}} \text{Acceptor}^{\overline{\bullet}} + R_3N^{+\bullet} \quad (1)$$

As a consequence of their low oxidation potentials, amines can serve as efficient electron donors to a wide variety of excited-state acceptors ranging from arene, cyanoarene, and arylalkene singlets to alkyl, α,β-unsaturated and aryl ketone triplets. The estimated ground- and excited-state energies and redox potentials of selected members of each of these acceptor families are given in Table 2.

Table 1. Oxidation Potentials $E_{1/2}(+)$ of Selected Amines, Amides, and Carbamates

Substrate	$E_{1/2}(+)$ (V)[a]	Substrate	$E_{1/2}(+)$ (V)[a]
Me_2NPh	+0.45[b]	$PhCH_2CH_2N(Me)CO_2Me$	+1.61[e]
Me_3N	+0.82[c]	$PhCH_2CH_2N(CH_2SiMe_3)CO_2Me$	+1.45[e]
Et_3N	+0.79[c]	MeNHPh	+0.25[d,f]
n-Bu_3N	+0.62[c]	Me_3SiCH_2NHPh	+0.10[d,f]
Me_2NCOMe	+1.82[b]	$Me_3SiCH_2CH_2NPh_2$	+0.30[d,f]
n-Bu_2NCO_2Me	+1.52[d,e]	$Me_3SiCH_2NPh_2$	+0.07[d,f]

Notes: [a]Versus Ag/AgNO₃ in MeCN.
[b]From Ref. 191.
[c]From Ref. 20.
[d]$E_{1/2}(+)$ versus SCE values were corrected by subtraction of 0.34 V.
[e]From Ref. 21.
[f]From Ref. 22.

Table 2. Reduction Potentials $E_{1/2}(-)$, Excited-State Energies $E_{0,0}$, and Excited-State Reduction Potentials $E_{1/2}^*(-)$ for Selected Arenes, Cyanoarenes, and Ketones

Substrate	$E_{\frac{1}{2}}(-)$ (V)	$E^1_{0,0}$ (eV)	$E^3_{0,0}$ (eV)	$E_{\frac{1}{2}}^{*1}(-)$ (V)	$E_{\frac{1}{2}}^{*3}(-)$ (V)
Benzene[a]	−2.74[b]	+4.76	+3.31	+2.02	+0.55
Naphthalene	−2.84[b,c]	+3.97[d]	+2.65[e]	+1.13	−0.19
Phenanthrene	−2.79[b,c]	+3.58[d]	+2.69	+0.79	−0.10
trans-Stilbene[a]	−2.60[b]	+3.53	+2.12	+0.93	−0.48
p-$(CN)_2C_6H_4$[a]	1.97[c]	+4.27	+3.03	+2.20	+1.06
1,2,4,5-$(CN)_4C_6H_2$[d]	−0.99	+3.83	—	+2.84	—
1,4-Dicyanonaphthalene[d]	−1.62	+3.45	—	+1.83	—
9,10-Dicyanoanthracene[d]	−1.23	+2.88	—	+1.65	—
Benzophenone	−2.06[b,c]	—	+2.97	—	+0.91
Benzil	−1.38[b,c]	+2.56	+2.32	+1.18	+0.94
Phthalimide	−1.81[b,c]	+3.46[f]	+2.98[f]	+1.65	+1.17
trans-MeCH=CHCOMe	−2.42[b,g]	—	+3.04[h]	—	+0.62
Cyclohexenone	−2.6[i]	—	+2.75[j]	—	+0.15

Notes: [a]Reduction potentials (versus Ag/AgNO₃) are corrected for those obtained versus SCE by substraction of 0.34 V.
[b]From Ref. 27.
[c]From Ref. 20, pp 41, 340, 184, 187.
[d]From Ref. 7, Vol. 6, pp. 237–238.
[e]From Ref. 28.
[f]From Ref. 29, p. 259.
[g]From Ref. 30.
[h]From Ref. 31.
[i]From Ref. 32.
[j]From Ref. 33.

2.2. Properties of Amine Cation Radicals

SET oxidations of amines leads to formation of amine cation (aminium) radicals, species with a rich history as transient intermediates in a host of interesting organic chemical processes. Since the structural, spectroscopic, and chemical properties of aminium radicals have been thoroughly discussed in an earlier and still informative review by Nelsen and his co-workers,[18] only a few pertinent aspects of the chemical properties of these intermediates will be outlined here. As would be expected for intermediates with high energy content and both radical and cationic character, aminium radicals participate in a number of fast and predictable reactions. For example, the N–H protons of primary and secondary amine cation radicals are acidic. The low pKa of 7 estimated for the Me_2NH cation radical as compared to 11 for its nonradical counterpart $Me_2NH_2^+$ exemplifies this general effect.[34] As a result of this property, one of the most characteristic, and often most efficient, reactions of these species is N–H deprotonation to produce aminyl radicals [Eq. (2)]. Ingold[35] has used a kinetic ESR technique to determine that the rate of the N–H proton transfer from the Me_2NH cation radical to solvents with H_O values ranging from –3 to –9 varies from 10^5–10^6 M^{-1} s^{-1}.

$$R_2\overset{\bullet+}{N}\text{–}H + B: \xrightarrow{-H^+} \overset{+}{B}\text{–}H + R_2N\bullet \tag{2}$$

Although lacking N-H protons, tertiary aminium radicals are also acidic species as a consequence of the presence of α-CH protons. α-CH deprotonation of these intermediates leads to generation of stabilized α-amino radicals.

$$R_2\overset{\bullet+}{N}CHR_2 + B: \xrightarrow{-H^+} \overset{+}{B}\text{–}H + R_2N\overset{\bullet}{C}R_2 \tag{3}$$

As a matter of fact, many tertiary amine oxidation processes are known to proceed by sequential SET–α-CH deprotonation pathways, leading mainly to products of oxidative dealkylation.

Owing to the importance of this α-CH deprotonation reaction, studies have been conducted recently to probe several features of tertiary amine cation radical acidity. In one effort by Das and von Sonntag,[36] both the pKa (ca. 8 in H_2O at 25 °C) and the second-order rate constant (7×10^8 M^{-1} s^{-1} in H_2O, 25 °C with Me_3N as base) for deprotonation of the Me_3N cation radical were determined by pulse radiolysis techniques. If acid–

base chemistry is indeed involved in the pathway leading to decay of this cation radical, the rates of non-aryl stabilized tertiary aminium radical deprotonation by strong (pKa *circa* 10) amine bases are larger than originally thought.[37]

The kinetics of α-CH deprotonation in the highly stabilized amine cation radical *N*-methyl-*N*, *N*-di-*p*-anisylaminium hexafluorarsinate (**1**) have recently been subjected to careful scrutiny in the stopped-flow experiments of Dinnocenzo and Banach.[38] These workers have found that the rate constants for deprotonation of this species in MeCN at 25 °C by isoquinuclidine bases with pKa values ranging from 15 to 20 are (1) 10^2–10^5 M^{-1} s^{-1}, (2) correlated with base strength with a β-value of 0.63 from a Brönsted plot, and (3) associated with small ΔH° (ca. 4 kcal/mol) and large negative ΔS° (ca. –22 cal/mol deg).

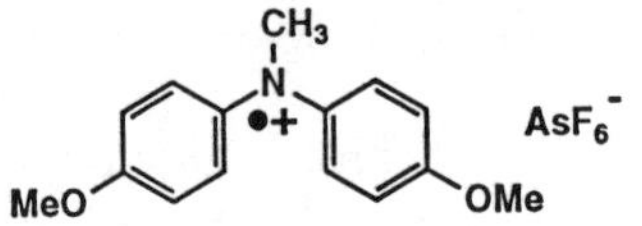

CH_3
N
•+
AsF_6^-
MeO
OMe

1

A derivative cyclic voltammetry method has also been employed to gain information about the kinetics of tertiary amine cation radical α-CH deprotonation. Using this technique, Parker[39] has measured the rate constants for proton transfer from *p*-substituted-*N,N*-dimethylanilinium radicals to pyridine and acetate as bases in MeCN. The data show predictable trends linking the rates with the ability of the *p* substituent to stabilize (lowering the rate) or destabilize (increasing the rate) the amine cation radical. For example, α-CH deprotonation rates were shown to vary from 1 to 2×10^5 with pyridine as the base and 4×10^6 to 3×10^9 with acetate as the base for the series of *p*-OCH_3 < *p*-CH_3 < *p*-Cl < *p*-CN < *p*-NO_2 *N,N*-dimethylanilinium radicals. Interestingly, these results indicate that amine cation radical α-CH deprotontation rates parallel their estimated[40] pKa values and, therefore, that the kinetic and thermodynamic acidities of these intermediates are linked. While these experimental results and their implications seem quite reasonable, the abnormally high (e.g., 22) kinetic deuterium isotope effect associated with these processes suggests that a degree of caution should be exercised with their interpretation.

Closely related to aminium radical α-CH deprotonation are other reactions where electrofugal groups are lost by α-heterolytic fragmentations [Eq. (4)]. One example of this general reaction type is the retro-aldol like cleavage [Eq. (5)] of cation radicals derived from tertiary β-amino alcohols.[41] Also, decarboxylations [Eq. (6)] of α-amino acid cation radicals[42] and demetallation of α-trialkylsilyl and -stannyl aminium radicals [Eq. (7)][23,24,26] represent commonly observed pathways open to these reactive intermediates. As in the case of α-CH deprotonations, the α-heterolytic fragmentation processes followed by aminium radicals are driven by a combination of factors including delocalization of the positive charge density into the α-C–C or α-C–MR_3 bonds and thermodynamic driving forces associated with formation of carbonyl π-bonds or metal–nucleophile σ-bonds.

$$R_2\overset{+\bullet}{N}-C(E)R_2 \xrightarrow{Y:} R_2N-\dot{C}R_2 + Y\text{-}E^+ \quad (4)$$

$$R_2\overset{+\bullet}{N}-CR_2-CR_2-O\text{-}H \xrightarrow{Y:} R_2N-\dot{C}R_2 + R_2C{=}O + Y\text{-}E^+ \quad (5)$$

$$R_2\overset{+\bullet}{N}-CR_2-CO_2H \xrightarrow{Y:} R_2N-\dot{C}R_2 + O{=}C{=}O + Y\text{-}E^+ \quad (6)$$

$$R_2\overset{+\bullet}{N}-C(MR_3)R_2 \xrightarrow{Y:} NR_2-\dot{C}R_2 + Y\text{-}MR_3^+ \quad (7)$$

(M = Si or Sn)

As electron-deficient odd-electron species, aminium radicals are also anticipated to display typical radical-like reactivity including H-atom abstraction and addition to electron-rich alkenes and arenes. The intramolecular variant of the H-atom abstraction process is seen in the familiar photoinduced Hofmann–Leffler–Freytag reaction transforming *N*-alkyl-*N*-chloroammonium cations to *N*-heterocyclic products.[43]

2.3. Aminyl and α-Amino Radicals

Neutral aminyl radicals are formed when primary or secondary aminium radicals undergo N–H deprotonation. While to our knowledge the chemistry of this class of reactive intermediates has not been fully explored, it should contain reactions characteristic of electron-deficient radicals. Thus, in the presence of alkyl radicals (radical pairs or diradicals), C–N bond formation should ensue. H-atom abstraction from alkyl chains and addition to electron-rich alkenes and arenes should also be representative of the chemistry of these intermediates.

Much more attention has been given to the study of the properties of α-amino radicals owing to their proposed intermediacy in a variety of tertiary amine oxidation reactions. The general pathway followed in these oxidations appears to involve SET from the amine donors to acceptor oxidants such as ClO_2,[44] ferricyanide,[45,46] and anodes,[47] followed by α-CH deprotonation. The α-amino radicals produced under these conditions are readily converted to iminium cations by a second SET step. The ready oxidation of α-amino radicals is a consequence of their much lower $E_{\frac{1}{2}}(+)$ values as compared to those of the tertiary amine precursors. The important photoelectrochemical measurement made by Griller and Wayner[48] show that these electron-rich odd-electron species have oxidation potentials in the range of –1.0 V and, as a result, they can serve as potent reducing agents. In this light, their reactivity with molecular oxygen through either radical addition or SET pathways and with readily reduced organic halides is easily understood.

α-Amino radicals are also known to react with alkenes [Eq. (8)].[49] The rates and regiochemistry of these addition reactions should be markedly influenced by substituents on the alkene moiety. The strong electron-donating nature of the amine function results in a large perturbation of the singly occupied molecular orbital (SOMO) energy in these species. As frontier orbital consideration suggests,[50] electron-withdrawing group-substituted alkenes (with low-energy LUMOs) should display higher reactivity with α-amino radicals. In addition, since electron-withdrawing group-substituted alkenes have larger β-coefficients in the LUMO, their reactions with these electron-rich radicals should display a high regiochemical preference for β-addition.

$$R_2N{-}\dot{C}R_2 \;+\; CH_2{=}CH{-}Y \longrightarrow R_2N{-}CR_2{-}CH_2{-}\dot{C}H{-}Y \qquad (8)$$

2.4. Summary

The excited-state reactions which serve as the focus of this review are promoted by SET from amine donors to a variety of organic acceptors. The transformations initiated in this way can be grouped according to the overall outcome of the processes. As shown in Scheme 2, the transformations can involve adduct or cyclization product formation, amine oxidative dealkylation, or acceptor reduction. In the sections

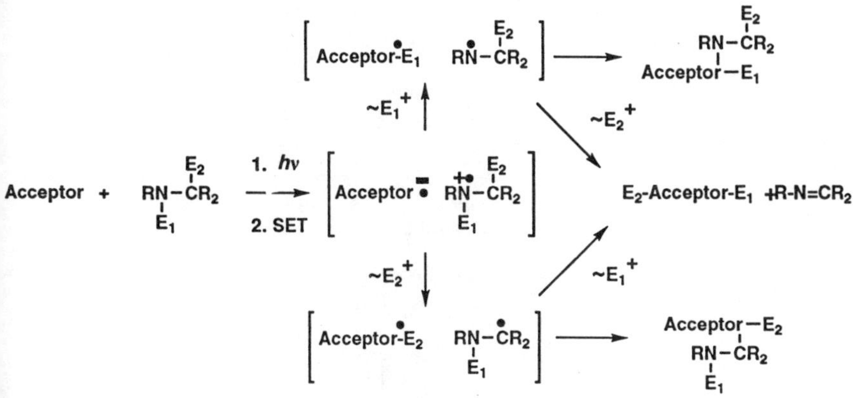

Scheme 2.

which follow, processes of these types will be described from the perspective of developing a picture of how the redox properties of amines, the chemical reactivity of intermediate aminium, aminyl, and α-amino radicals, and the properties of acceptors combine to govern the often mechanically complex yet synthetically unique photoinduced SET chemistry of amines.

3. SET PHOTOCHEMISTRY OF AMINES DRIVEN BY AMINIUM RADICAL N–H OR α-CH DEPROTONATION

3.1. Arene–Amine SET Photochemistry

As anticipated based on excited- and ground-state redox properties, the photochemistry of arene acceptor–amine donor systems is dominated by reaction pathways initiated by SET.[51] Early examples of this type of reactivity were uncovered by Bryce-Smith and his co-workers in their studies of primary and secondary amine photoadditions to benzene. Thus, the formation of the cyclohexadienyl amines (e.g., **2**) by irradiation of benzene solutions containing piperidine and cyclohexylamine, respectively, is representative of the SET-promoted reactions of these systems.[52] In a related manner, benzene was observed to photoadd to pyrrole to generate the adduct **3** resulting from α-C rather than N bonding.[53]

In a later study of this photochemistry, Bryce-Smith thoroughly analyzed the addition reactions of a variety of primary and secondary amines to benzene.[54] The reactions were shown to occur by way of the lowest energy benzene-singlet excited state and to generate amine-substituted 1,3- and 1,4-cyclohexadienes along with bicyclo[3.1.0]hexenyl and hexatrienyl amines. For example, irradiation of a deoxygenated solution of *t*-butylamine in benzene gives rise to the adducts **4**–**7**. Importantly, these processes do not occur in polar protic solvents like MeOH, but rather are favored in nonpolar aprotic media (e.g., C_6H_6, Et_2O, cyclohexane). In addition, the bicyclic and hexatrienyl amines **6** and **7** were found to be secondary photoproducts arising by familiar 1,3-cyclohexadiene excited-state processes.

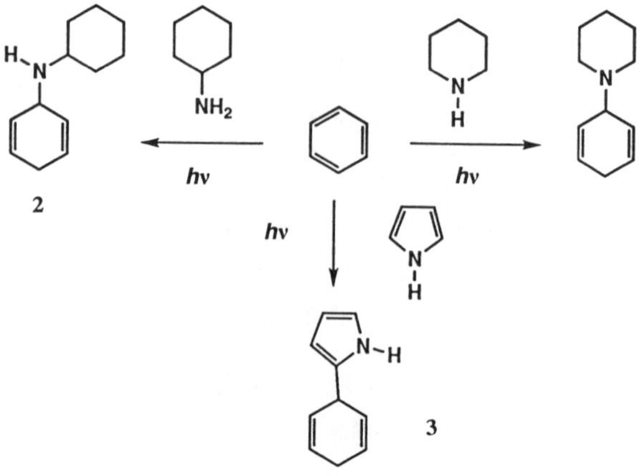

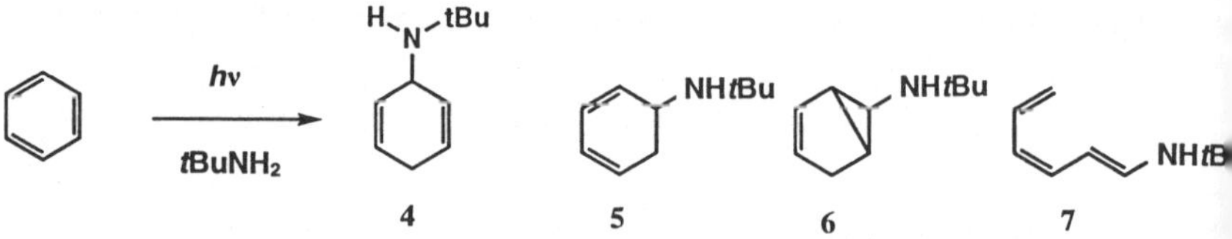

The contrast between both the nature of the photoproducts formed in these primary- and secondary-amine photoreactions and the effects of polar solvents on them, on the one hand, and on the other hand those associated with the tertiary amine additions to excited arenes is informative. Thus, while tertiary amines such as Et_3N add to benzene by way of its singlet excited state, the efficiencies for generation of α-adducts **8** and bicyclic tetraenes **9** are enhanced in proceeding from solvents of low to high polarity and greater when the protic nature of the medium is high (e.g., MeOH-H_2O).[44]

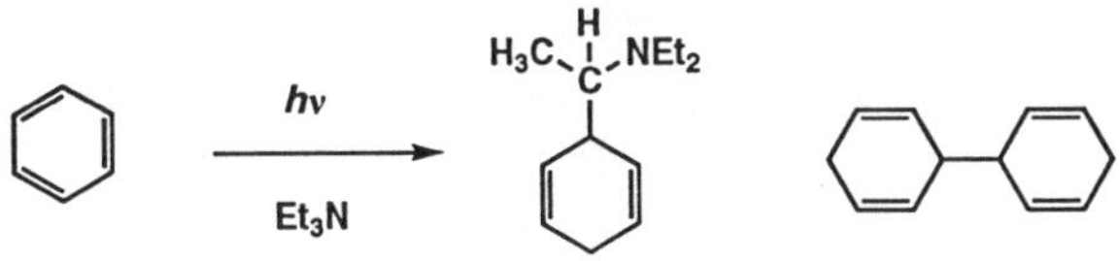

8 9

The mechanistic consequences of these observations are further clarified by the results of studies conducted by Yang and Libman[55] probing amine photoreactions with anthracene. Similar solvent effects on the efficiencies of secondary and tertiary amine photoadditions to this arene were noted. Importantly, it was shown that the solvent effect does not originate in the excited-state SET step, since all amines quench anthracene fluorescence at a higher rate in MeCN than in C_6H_6. Also, it was found that photoreaction of Et_2ND with anthracene results in production of **10–12**, all containing deuterium labels at the dihydroanthracene 10-position.

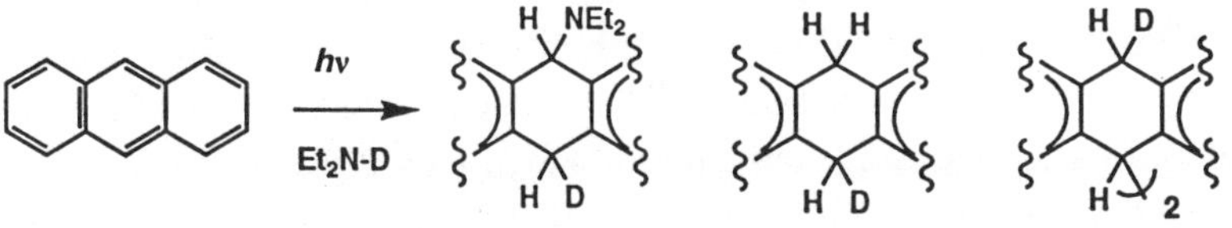

10 11 12

The combined results of these investigations suggest that primary and secondary amines photoadd to arenes by a process initiated by SET. Exciplex intermediates arising in this way in nonpolar media undergo proton transfer faster than they collapse to free ion radicals. The resulting solvent-caged radical pairs then couple by C–N bond formation to yield amine–arene adducts (Scheme 3):

The detailed sequence for tertiary amine additions to arenes has some features in common with that for the primary and secondary counterparts. That these additions are induced by SET to the singlet arene is evidenced by the Yang and Libman[55] observations that fluorescence quenching rates depend inversely on tertiary amine ionization potentials and directly on solvent polarity. In these cases, however, the quantum efficiencies for product formation are enhanced by increases in solvent polarity. Furthermore, photoaddition of Et_3N to benzene in CD_3OD results in production of the diastereomeric mono-D adducts.[54] Clearly, the mechanistic differences between the tertiary and the primary and secondary

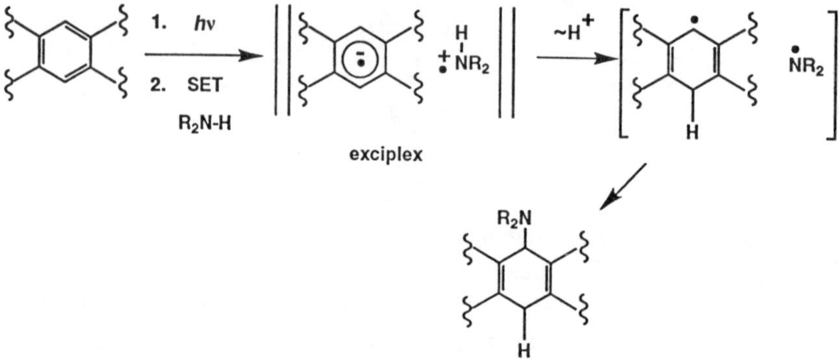

Scheme 3.

amine photoreactions with arenes must lie in the chemistry of the ion radical intermediates formed in the initial SET events. It appears that in the absence of a fast N–H proton transfer process in the singlet exciplex formed between tertiary amines and arenes, decay by electron backtransfer (BSET) dominates (Scheme 4). However, in polar media, exciplex decay to free radical ion intermediates can compete with BSET. Deprotonation of the free amine radical cation then occurs in a manner mediated by solvent so as to account for the D-labeling result summarized above. It is important to emphasize that this interpretation suggests that N–H proton transfer from primary or secondary amine cation radicals in exciplexes (or contact ion radical pairs) is much faster than α-CH proton transfer in related tertiary amine derived exciplexes. This conclusion is in accord with results of direct measurements of amine cation radical N–H and α-CH deprotonation rates and their thermodynamic (pKa) acidities.

Studies of the SET-induced photochemistry of arene–amine systems have led to a number of additional observations that add information about the detailed mechanical nature of these processes. For example, Pac and Sakurai[56] and Davidson[57] have independently explored the photoadditions of tertiary amines to anthracene. The former group found that the 4+4 dimer **13** is the major product (90%) of the anthracene photoreaction with dimethylaniline in C_6H_6, while **13** along with the reduction products **14** and **15** and mainly adduct **16** (65%) are produced when the more polar MeCN is used as solvent. D-labeling experiments were employed to demonstrate that the reaction in MeCN proceeds through the intermediacy of free ion radicals (Scheme 4). The exclusive

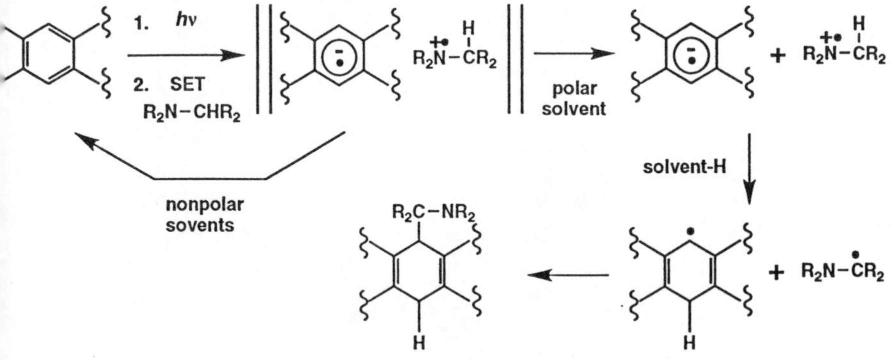

Scheme 4.

formation of dimer **13**, a substance also obtained by direct irradiation of anthracene in the absence of amines, suggests that the exclusive pathway open to the aniline–anthracene exciplex is decay to ground-state reactants. A similar observation and interpretation were made by Davidson,[57] and intramolecular examples of this type of reactivity have been documented.[58]

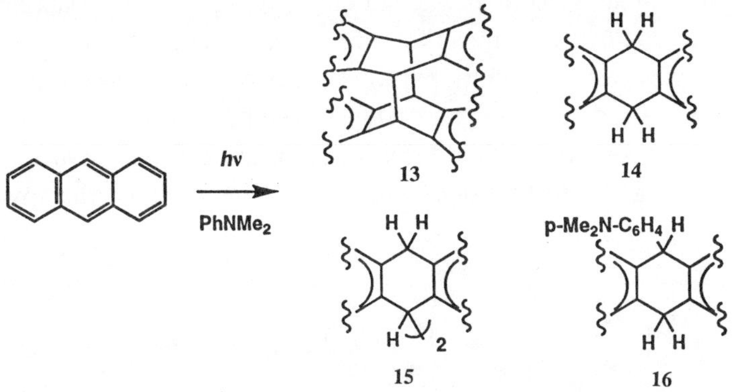

A discussion of tertiary amine–arene SET-induced photochemistry is required to include a summary of the interesting seminal work of Baltrop and his co-workers[59,60] on the reactions of naphthalene and its derivatives. These researchers found that irradiation of naphthalene and its methoxy-substituted analogs in the presence of Et_3N generates the Birch-type reduction products **17** and **18** along with dimers **19** and amine–arene adducts **20**. Deuterium-labeling experiments (e.g., irradiation in D_2O–MeCN) demonstrate that the formed 1,4-dihydronaphthalenes derive both their new hydrogens from solvent. These results, in combination with those from emission spectroscopy experiments, led Baltrop to the reasonable conclusion that photoreactions in these systems are initiated by SET from the amine to the singlet-excited naphthalenes and are propagated by arene anion radical protonation, hydroarenyl radical reduction, and hydroarenyl anion protonation sequences (Scheme 5).

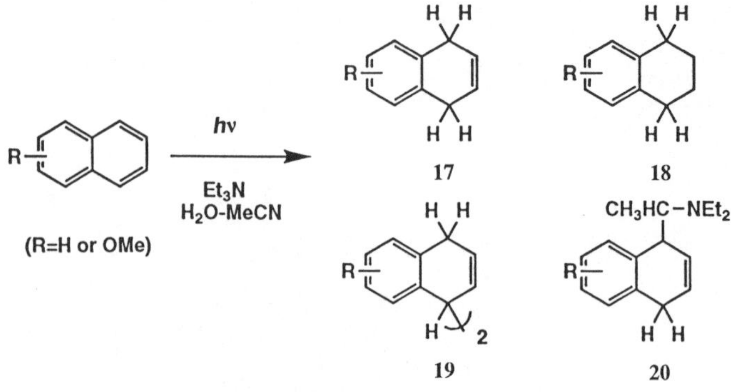

The closely related SET-induced photosubstitution chemistry of N-heteroarenes with amines is worthy of mention at this point since the mechanical pathways involved and the factors governing reaction type appear to contrast with those for amine–arene photoadditions. As with simple hydrocarbon analogs, pyridine reacts with secondary and tertiary amines on irradiation to yield adducts resulting from bonding at the amine α-carbons. Examples of this behavior are present in investigations by Gilbert and Krestonosich,[61] in which it was shown that pyridine

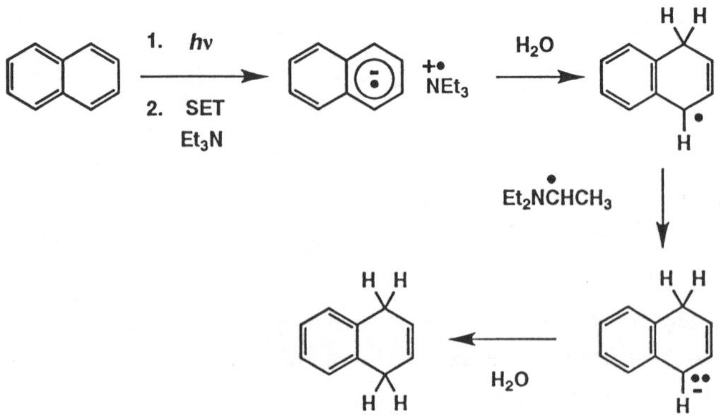

Scheme 5.

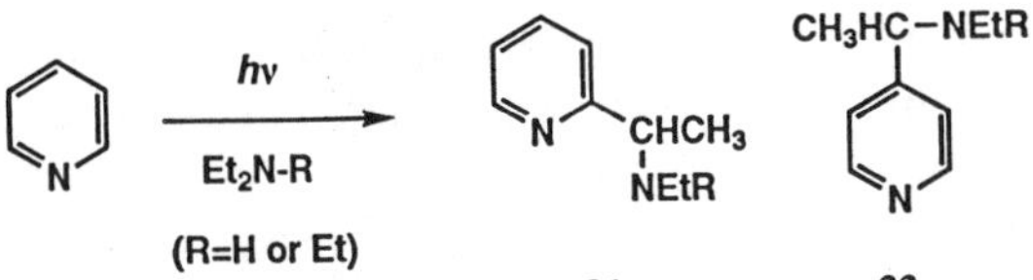

photochemically adds Et_2NH and Et_3N to produce the 2- and 4-substituted aminoalkyl adducts **21** and **22**. Consistent with the exclusive formation of αC- rather than N-adducts in the reaction of Et_2NH is the observation that irradiation of a pyridine solution containing tBuNH$_2$ does not lead to substitution-product formation. This contrasts with the photoaddition of this amine to benzene, which yields 1,2- and 1,4-adducts (see above). On the other hand, primary and secondary amines (e.g., tBuNH$_2$ and Et_2NH) do react photochemically with 2-fluoropyridine to give substitution product **23** arising by N–C bond formation.[61] But the related photoreactions of these amines with fluorbenzene generate complex mixtures containing both simple substitution product (**24**) as well as those of *ortho, meta*, and *para* addition (**25–27**).[62,63]

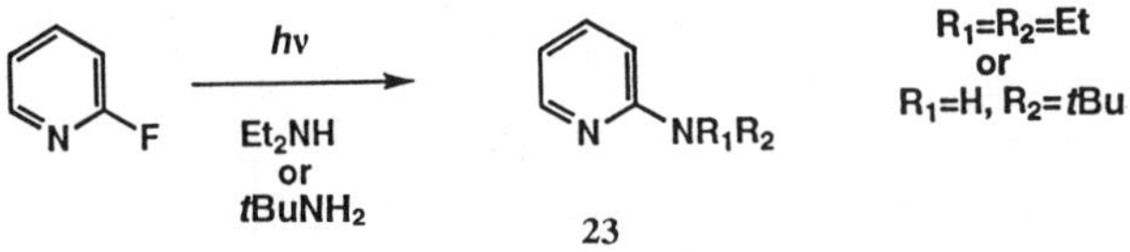

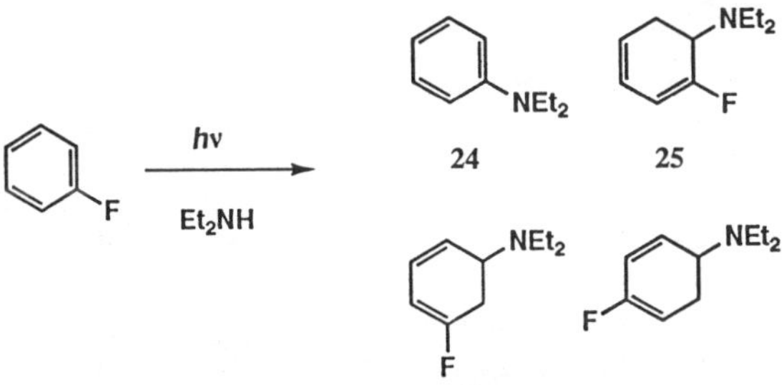

The different pathways followed in the arene versus N-heteroarene photoreactions with amines could be a consequence of the operation of different mechanisms. In this light, Gilbert[61] suggested that, unlike the arene processes, the pyridine reactions follow H-atom abstraction rather than SET routes. According to his reasoning, H-atom abstraction by the pyridine n-Π* excited state from amines occurs at the α-CH rather than N–H position to produce α-amino radicals, which couple to the hydropyridinyl radical to produce dihydropyridine precursors of the adducts. However, this mechanism is unsuitable in explaining the nature of the Et_2NH fluoropyridine photoreaction. In that case, Gilbert invoked a conventional photonucleophilic displacement pathway. These interpretations should be judged with a degree of skepticism since SET from amines to the excited state of N-heteroaromatic substances should be more facile (higher reduction potentials and similar excited-state energies) than those to the related hydrocarbon acceptors.

It is likely that SET is involved as the initial step in the N-heteroarene reactions with amines and that the differences in the nature of the reaction pathways followed are due to the properties of the anion radical intermediates or photoaddition products resulting from these substances. For example, it is possible that N–C bonding in photoreactions of primary and secondary amines with pyridine does indeed occur but that it is reversible owing to the expected high reactivity of the 2- or 4-amino 1,2- or 1,4-dihyropyridines, **28** and **29**, produced. Thus, the observation that only αC-addition products are formed might simply be a result of the stability of these adducts as compared to those arising by N-C bond formation, which rapidly revert to starting materials. The dihydropyridine adduct **30** derived by SET-promoted N-C bonding in the photoreactions of primary and secondary amines with fluoropyridine should likewise be unstable. However, in these cases elimination of H-F would be competitive with return to starting materials and would lead to observed substitution products.

Recently, Caronna and his collaborators[64,65] have observed closely related photoaddition processes in which primary and secondary amines add to pyridine-*bis*-carbonitriles to form aminopyridine products (e.g., **31** → **32** + **33**). Interestingly, these photosubstitution reactions were explained by using SET mechanisms.

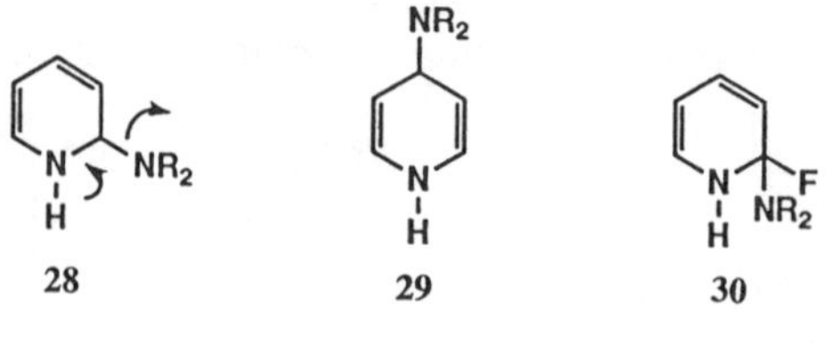

28 29 30

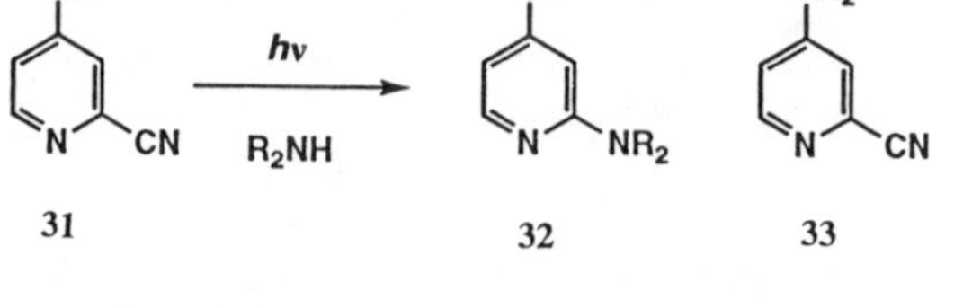

31 32 33

3.2. Halo- and Cyano-Arene Photosubstitution Reactions with Amines

SET-promoted photoreactions of halo- or cyano-arenes with amines are characterized by the formation of both substitution products having amino- or aminoalkyl-arene structures and reduction products in which hydrogen is substituted for the halogen or cyano function (Scheme 6). In general, several mechanisms are possible for these processes. One of these follows a sequence involving (1) SET-promoted formation of ion radical pairs **34**, (2) N–H or α-CH proton transfer, (3) N–C or C–C bond formation, and (4) elimination of either HX or HCN (Scheme 6). An alternative route must be considered owing to the high propensity for loss of nucleofugal halide and cyanide anions from the halo- and cyano-arene anion radical intermediates. This process generates aryl radicals that can serve as precursors for both adduct and reduction-product formation. Finally, photohomolysis of the C–X bond could serve as a direct route for aryl radical formation in the excited-state chemistry of some aryl halides.

3.2.1. Haloarene–Amine Photochemistry

From the time of the initial observations of amine–aryl halide photochemical reactivity, a number of studies have been conducted to define the mechanistic hierarchy for these processes and how it is altered by variations in the haloarene, amine, and medium. Among the earliest observations made in this area were those by Pac and his co-workers[66] in their pre-1972 investigations of *N,N*-dimethylaniline photoreactions

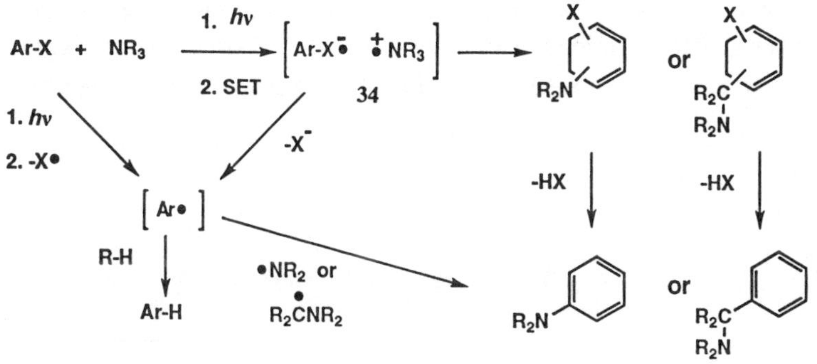

Ar-X + NR3 — 1. hν, 2. SET → [Ar-X•⁻ •⁺NR3] 34

1. hν
2. -X•
-X⁻
[Ar•]
R-H
Ar-H
•NR2 or
R2ĊNR2
X
R2N
or
R2C
R2N
-HX
-HX
R2N
or
R2C
R2N

Scheme 6.

with halobenzenes. These workers found that irradiation of dimethylaniline in methanol solution and containing triethylamine and chloro-, bromo-, or iodo-benzene results in formation of the triethylammonium halide in quantitative yields along with the dealkylation, reduction, and addition products shown in Scheme 7. The ratio of adduct **36** to the reduction product, benzene, was found to vary greatly in the series PhCl (1.8) > PhBr (1.1) > PhI (0.2). Likewise, the extent of D-incorporation in PhH, produced by reactions in which $(CD_3)_2$ NPh was used, is also highly dependent on the halogen (66% with PhCl, 42% with PhBr, and 12% with PhI). An interesting interpretation of these results is found in Pac's suggestion that the photoreactions occurring from the aniline excited-singlet state proceed through either exciplex intermediates arising by SET to PhCl and PhBr or free ion radicals when PhI is the acceptor. Thus, benzene produced by the pathway involving collapse of the free PhI anion radical arises by H-atom abstraction by the phenyl radical mainly from solvent. In contrast, phenyl radicals produced by halide loss in contact ion radical pairs $[DMA^{+\cdot}\ PhX^{-\cdot}]$ abstract hydrogen from the neighboring DMA cation radical.

In a later effort, Bunce and Gallagher[67] re-explored the photochemistry of arylamine–haloarene systems promoted by irradiation of the amine. This work was stimulated by a general interest in developing a detailed understanding of the mechanistic basis of sensitization methods for photodestruction of aryl halides, which are often environmental pollutants. Prior investigations by Miller and Narang[68] had shown, for example, that diethylamine sensitizes the photodestruction of DDT. Bunce and Gallagher's studies focused on the dehalogenation reactions of chlorinated biphenyls promoted by irradiation of dimethylaniline. Consistent with the earlier observations of Pac,[66] the chlorobiphenyl photoreactions were noted to derive mainly from the amine singlet and to have greater efficiencies in polar (e.g., MeOH) than in nonpolar (e.g., cyclohexane) solvents. Interestingly, an attempt to uncover longer wave-

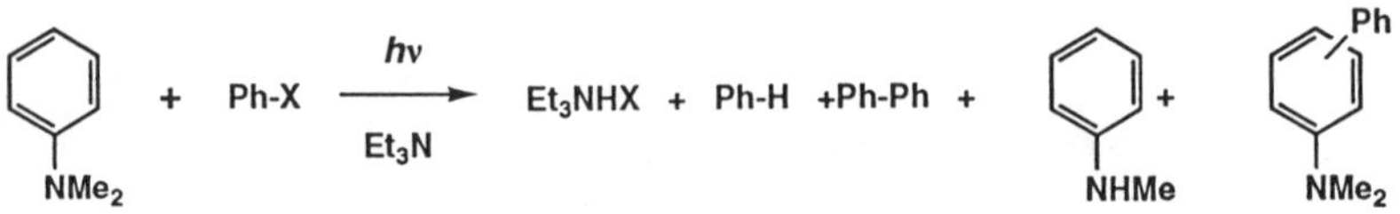

NMe₂ + Ph-X —hν, Et₃N→ Et₃NHX + Ph-H + Ph-Ph + NHMe (35) + Ph, NMe₂ (36)

Scheme 7.

length-absorbing amines as photosensitizers of the chlorobiphenyl reactions was unsuccessful. For example, both N-methylcarbazole and acridine were not effective in promoting the SET-induced dehalogenation reactions, presumably because of their insufficiently low excited-state reduction potentials.

SET-promoted photodehalogenation reactions of haloarenes can also be promoted by direct irradiation in the presence of nonabsorbing aryl or alkyl amines. An example of this is found in the work of Kuzmin,[69] which showed that anilines quench the fluorescence of 9,10-dichloro- and 9,10-dibromoanthracene with rate constants that increase with decreasing amine ionization potential (i.e., $PhNH_2 < PhNHMe < PhNMe_2$). Moreover, from the dependencies of fluorescence quenching and reaction quantum yields on aniline concentrations these workers concluded that the photodehalogenation reactions occur from the haloarene singlet excited states. Finally, fluorescent exciplexes were detected when both the dichloro- and dibromo-anthracenes were irradiated in heptane solutions containing $PhNMe_2$.

A number of efforts in this area have concentrated on the photochemistry of halogenated biphenyls, owing to the high environmental levels of these substances and the desire to develop cheap, efficient, and nonpolluting methods for their removal. In this area, Ohashi and his co-workers[70] explored the photodehalogenation of 4-chlorobiphenyl (**37** → **38**) induced by direct irradiation in the presence and absence of amines. The results of this effort led to conclusions similar to those found by Kuzmin.[69] Tertiary amines were found to be superior at facilitating the dehalogenation, as were polar solvents (e.g., at $[Et_3N] = 0.15$ M, $\phi = 0.49$ in MeCN, 0.12 in *t*BuOH, and 0.08 in cyclohexane). The combined observations led Ohashi to propose that the photoconversion of **37** to **38** occurs from the singlet of **37** and involves SET from the amine to generate the chlorobiphenyl anion radical. Chloride anion loss followed by H-atom transfer to the formed aryl radical completes the pathway.

A number[71–75] of detailed investigations of the mechanisms of haloarene–alkylamine photoreduction reactions have been conducted since the time of these initial efforts. One of the more informative of the recent studies, conducted by Davidson and Goodin,[75] demonstrated that the quantum yield enhancements by alkylamines of these photoreductions depend on the nature of the substrate, the halogen substituent, and the solvent. For example, irradiation of MeOH solutions of 4-bromoanisole and related bromo-aromatics leads to dehalogenation reactions that are not made more efficient by the addition of Et_3N. In contrast, Et_3N has an accelerating influence on corresponding chloroarene photodehalogenations. Thus, it appears that at least two mechanistic routes can operate in these reactions, one involving SET from the amine to the haloarene and the other a direct excited-state photohomolysis. The latter pathway predominates in haloarenes with weak C-X bonds and where rapid homolysis results in short-lived (nonquenchable by SET) singlet excited states.

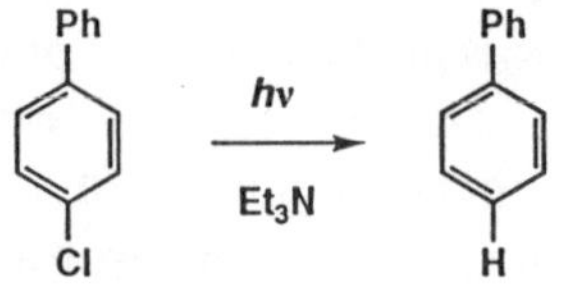

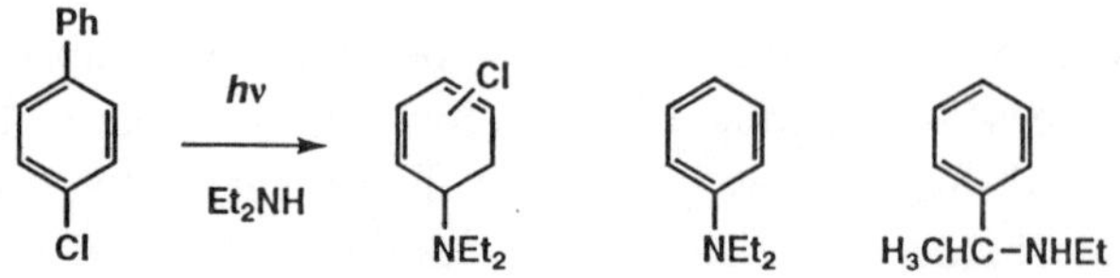

As pointed out earlier in this review, SET-induced photoreactions between amines and haloarenes also result in the production of adducts arising from amine N- and αC-bonding. Representative examples of this reactivity pattern are found in the observations made by Gilbert and his co-workers.[76] Thus, irradiation of chlorobenzene in the presence of Et_2NH gives the N-addition and N- and αC-substitution products **39–41**, respectively.

3.2.2. *Cyanoarene–Amine Photochemistry*

Cyanoarenes have played an important role in the development of SET photochemistry. Special attention has been given to the highly conjugated, long-wavelength-absorbing members of this series (e.g., 9,10-dicyano and 2,6,9,10-tetracyanoanthracene) owing to their usefulness as SET-sensitizers in a variety of photochemical processes.[77] As SET-sensitizers, their singlet excited states can be selectively populated by long-wavelength excitation ($\lambda > 320$ nm) and can easily be reduced by donor substrates to produce cyanoarene anion radicals and substrate cation radicals. Importantly, in this role the cyanoarenes must have stable anion radicals to prevent their participation in the ensuing chemistry until the penultimate step, when electron backtransfer occurs to form the cyanoarene ground state. The sequence involved in this chemistry is best exemplified by the 9,10-dicyanoanthracene (DCA)-sensitized methanol addition to alkenes observed and elegantly explored by Arnold and his co-workers[78] (Scheme 8):

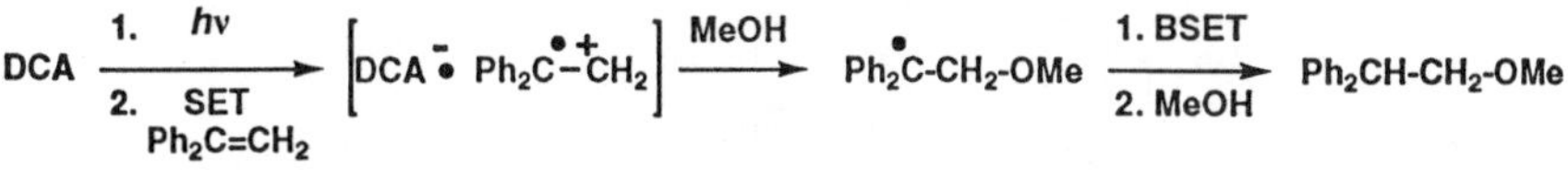

$$\text{DCA} \xrightarrow[\text{2. SET, } Ph_2C{=}CH_2]{\text{1. } h\nu} \left[\text{DCA}^{\bullet -}\; Ph_2\dot{C}\text{-}\overset{+}{C}H_2\right] \xrightarrow{\text{MeOH}} Ph_2\dot{C}\text{-}CH_2\text{-OMe} \xrightarrow[\text{2. MeOH}]{\text{1. BSET}} Ph_2CH\text{-}CH_2\text{-OMe}$$

Scheme 8.

Clearly, the anion radicals derived from the highly conjugated cyanoarenes must have great stability (long lifetimes) in order to ensure their survival in the presence of acidic, radical, and cationic species. A recent study probing this issue was conducted by Whitten and his co-workers.[79] They observed that when a deoxygenated solution of DCA containing Et_3N and $n Bu_4N\ (PO_4H_2)$ is irradiated, the DCA anion radical is both efficiently (91%) formed and stable for a period of weeks even in the presence of water.

Unlike DCA, which is relatively (see below) inert in SET-induced photoreactions, less-conjugated cyanoarenes participate in efficient photosubstitution reactions with amines. Key observations in this area have come principally from the laboratory of Ohashi and his collaborators.[80,81] Both *p*- and *o*-dicyanobenzene, for example, participate in high yielding (ca. 80%) photoreactions with Et_3N to give products of substitution and reduction **42–44**. These processes most probably proceed by a sequence in which excited-state SET is followed by α-CH deprotonation (with amine or cyanoarene anion radical as base) and C-C bonding to produce the dihydroarene adduct. Elimination of HCN then affords the final adduct **42**. The origin of the alkylation product **43** in these photoreactions is also of interest. Ohashi has shown that these substances arise by secondary photoreactions of the initially formed adduct **42**. Several mechanisms for the conversion of **42** to **43** can be envisaged, including those portrayed in Scheme 9 involving photoreductive benzylic-cleavage or intramolecular SET-induced fragmentation. It is interesting that tetraal-

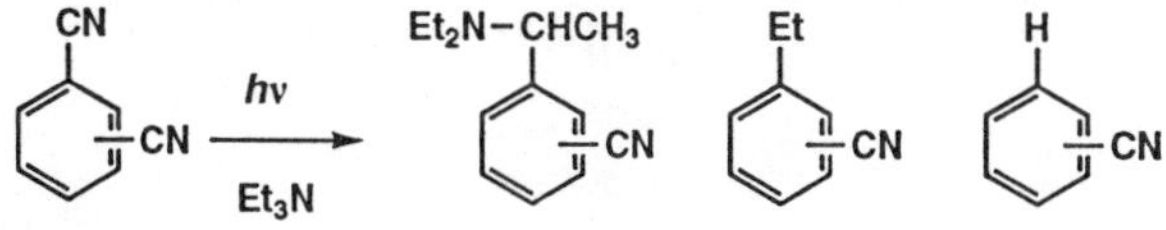

CN, CN (*o* and *p*) $\xrightarrow[Et_3N]{h\nu}$ **42** ($Et_2N-CHCH_3$, CN) + **43** (Et, CN) + **44** (H, CN)

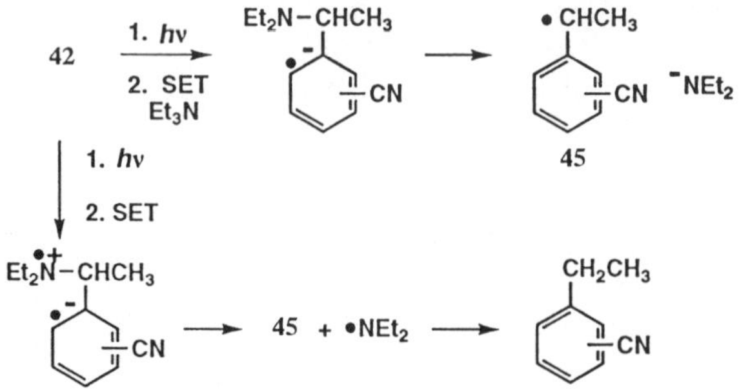

Scheme 9.

kyl ureas also undergo related SET photoreactions with cyanoarenes to produce similar distributions of products (e.g., **46** → **47** + **48**).[82]

Ohashi's studies of amine–cyanoarene photochemistry have uncovered a remarkable and unprecedented reaction.[83] The process involves the primary, secondary, and tertiary amine–promoted photoconversion of DCA in MeCN to its 9-amino-10-cyano analog **50** in high yields (60–95%). Information about the mechanistic details of this reaction has come from the observation that (1) the reaction efficiencies are enhanced by the presence of H_2O in the MeCN solvent system, (2) the amino group in **50** comes from the nitrogen of MeCN, and (3) acetaldimine **49** is an intermediate in this process. Thus, it appears that SET from amines to DCA^{S_1} initiates a cascade of events (Scheme 10) leading to generation of the hydroanthracenyl anion **51**, which then reacts with the nitrile function of the solvent to yield **49**.

It is intriguing that DCA is unlike other cyanoarenes not because of the stability of its anion radical, as is understandable, but rather because of the unique nature of the SET-promoted photoreaction in which it participates. Additional information about this photochemistry has come from more recent studies by Mariano, Yoon, and their co-workers.[84] Irradiation of DCA in an MeCN solution containing Et_2NMe

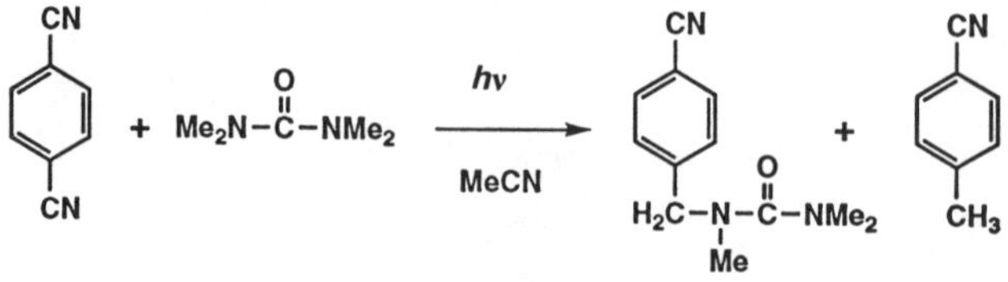

46 **47** (19%) **48** (18%)

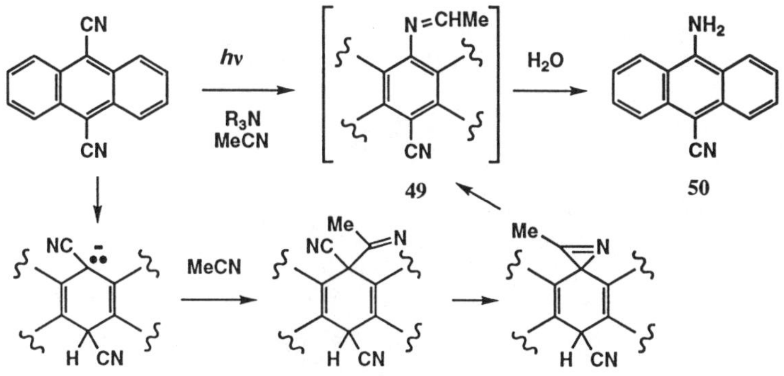

Scheme 10.

was found to produce mainly (41%) the aminoanthracene **50** along with lesser (yet finite) quantities of the photoaddition derived products **52** (2%) and **53** (4%). Surprisingly, the photoreaction of DCA in the presence of the silyl analog, Et_2NCH_2TMS, while giving the same products, favors formation of **52** (19%) and **53** (5%) to a greater extent as compared to **50** (29%).

The origin of the differences in the nature of the excited-state reactions followed by DCA as compared to less-delocalized cyanoarenes such as 1,4-dicyanonaphthalene and the dicyanobenzenes with tertiary amines is not easily explained. The recent observations that the DCA anion radical is long lived[79] relative to other cyanoarenes[85] suggests that this reactive intermediate should resist reactions with radical and cationic intermediates as well as acids. Rough estimates of the pKa of the protonated DCA anion radical suggest that it falls in the region <<O.[86] When taken together, these observations and estimates show why DCA might be unique in this series as an SET-sensitizer.[87] In addition, the low basicity of the DCA anion radical perhaps explains why it is not protonated by amine cation radicals that have either acidic NH or αC–H protons and, consequently, why it could be available as a reactive nucleophile in the aminoanthracene **50**–forming process observed by Ohashi.

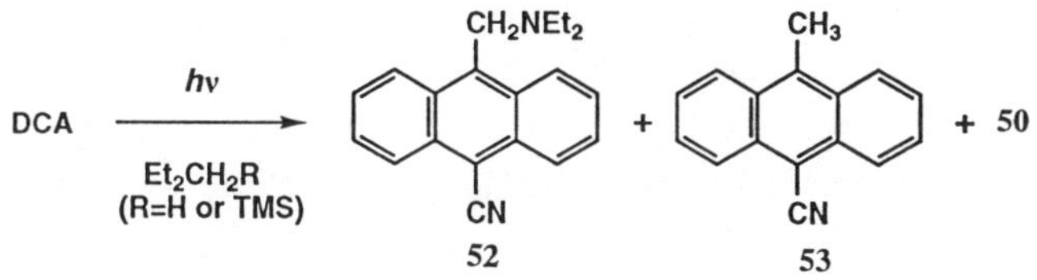

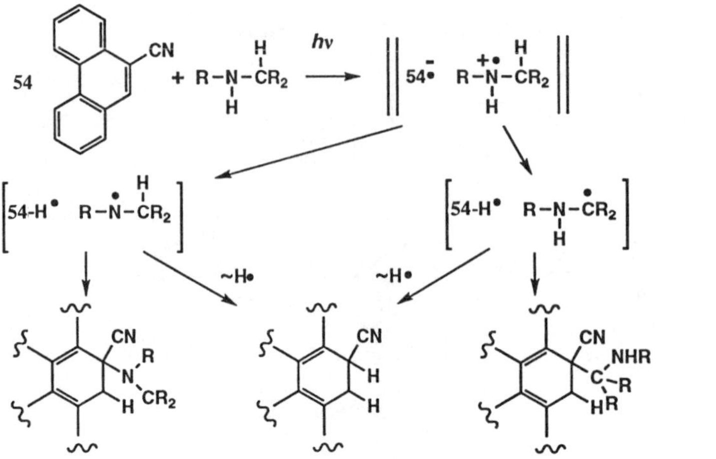

Scheme 11.

Information about the competition between N–H and αC–H deprotonation of amine cation radicals in routes for primary and secondary amine additions to cyanoarenes has come from the extraordinarily thorough studies of Lewis and his collaborators[27,88,89] of the photochemistry of 9-cyanophenanthrene **54**. The pathways for N- and αC-adduct and reduction product formation (Scheme 11) from **54** were elucidated in this effort by use of N–D and αC–D labeling experiments. Partitioning of the amine cation radical–cyanophenanthrene anion radical contact pairs (i.e., exciplex) by N–H and αC–H proton transfer was found to be governed by two factors: (1) α-substituents on the amine, which through their control of C–H bond dissociation energies effect the rates of cation radical αC–H deprotonation; and (2) solvent polarity, an increase in which slows N–H deprotonation.

The photoaddition of tertiary urea **55** to cyanophenanthrene **54** resembles closely the tertiary amine reactions with **54** with one curious but telling exception.[90] Just as with the tertiary amine photoreactions, the major products obtained by irradiation of **54** in the presence of **55** are an adduct **56** and a reduction product **57**. However, the regiochemistry for adduct formation in the urea reaction is opposite to that for the amine additions to **54**. Another major difference is that photoreaction of cyanophenanthrene with the urea occurs from the triplet excited state of the arene since **55** does not quench the fluorescence intensity or reduce the singlet lifetime of **54**. Thus, it is likely that a radical addition rather than radical coupling mechanism is operable in the urea reaction. Formation of the α-ureidomethyl radical **58** in this pathway could occur by H-atom abstraction by $\mathbf{54}^{T_1}$, as suggested, or, as we believe, by deprotonation of free urea cation radicals formed by triplet ion radical pair collapse. Addition of **58** to ground state **54** would occur in the observed regiochemical manner as predicted by FMO considerations. SET from the anion radical of **54** to the formed α-cyano radical followed by protonation would then generate the observed adduct **57**.

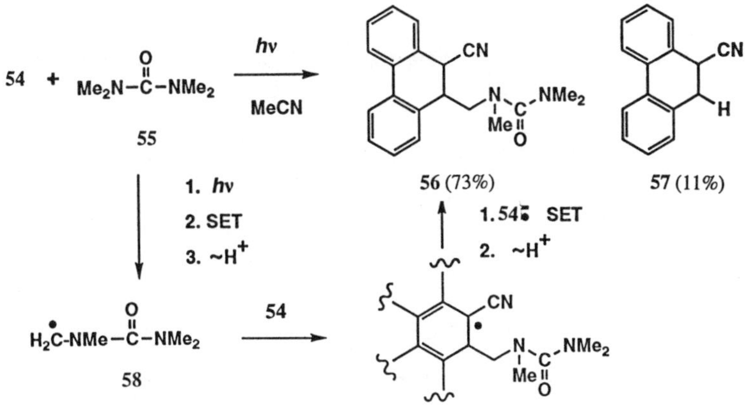

3.3. Photoadditions of Amines to Styrenes and Their Derivatives

Among the first of several observations of amine photoaddition reactions with phenyl-conjugated alkenes were those made independently by Kawanisi and Matsunaga[91] and by Cookson and his co-workers[92] during the early 1970s when SET mechanisms were less "in vogue." Studies by the Japanese group[91] at that time showed that secondary amines (e.g., morpholine, pyrrolidine) participate in reactions with diphenylacetylene **59** to generate products of photoreduction and photoreductive animation. This chemistry is exemplified by the formation of diphenylethane **60**, its amino analog **61**, and deoxybenzoin **62** by irradiation of **59** in neat pyrrolidine (Scheme 12). A generalized scheme for these transformations involves the intermediacy of stilbene **63**, serving as the precursor of **60** and **62**, and the enamine **64**, the hydrolysis of which gives ketone **62**.

In pioneering work by the Cookson group,[92] interesting photoaddition reactions of tertiary amines to styrene derivatives were uncovered. Examples of these modestly efficient reactions are given in Scheme 13

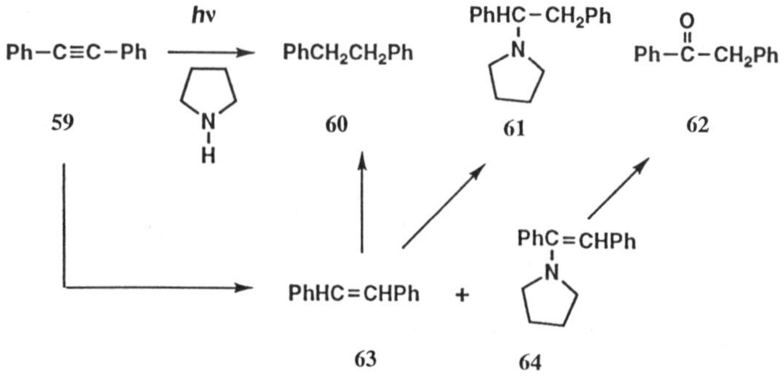

Scheme 12.

along with processes showing competitive N–C and αC–C bond-forming routes for photoreactions of primary amines with these arylalkenes.

Cookson's exploratory investigations stimulated later detailed and enlightening studies of these photoreactions by Lewis and his co-workers. These efforts have provided a more thorough picture of the mechanistic complexity of the photoadditions as well as insight into the factors governing the relative efficiencies of N–C and αC–C adduct formation and the αC-regioselectivities for additions of unsymmetrically substituted tertiary amines. An excellent survey of Lewis' studies can be found in an earlier review.[93] Consequently, only a brief summary of the salient aspects of this chemistry will be given here. Initial information about the

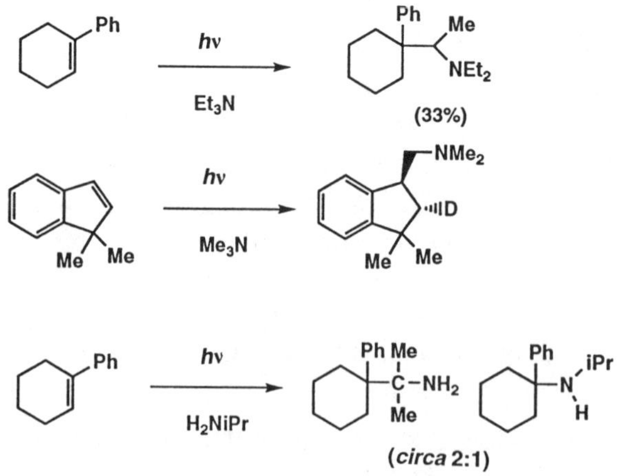

Scheme 13.

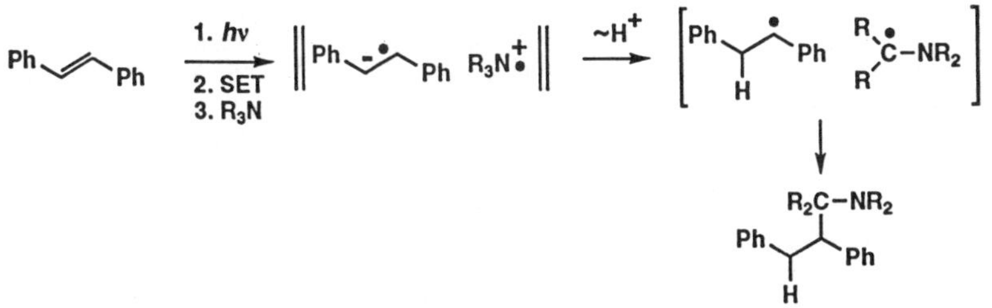

Scheme 14.

SET nature of the amine–arylalkene photoaddition processes came from investigations of solvent effects on tertiary amine–stilbene reactions. Specifically, the fact that the relative efficiencies for fluorescent exciplex and photoadduct formation vary inversely and directly, respectively, with increases in solvent polarity is the characteristic signature of photoadditions initiated by SET from the amines to the stilbene singlet excited state. The contact ion radical intermediates generated in this fashion undergo either proton transfer to yield radical precursors of products or light emission depending on the solvent (Scheme 14).

As expected on the basis of the sequence given in Scheme 14, the rate constants for tertiary amine quenching of stilbene fluorescence are inversely dependent on amine ionization potentials. Secondary and primary amines also photoadd to stilbene singlets, this time yielding adducts resulting from N–C rather than αC–C bonding. The higher thermodynamic acidity of N-H as compared to αC–H protons in amine cation radicals apparently translates into a higher kinetic acidity in the ion–radical pairs that serve as intermediates in these reactions.

A broadly interesting result obtained from Lewis' studies of tertiary amine photoadditions to stilbenes arose from efforts probing the regiochemical course of unsymmetric amine reactions. Lewis recognized that the product ratios in these processes (Scheme 15) provide information about how substituents can influence the kinetic αC–H acidities in amine cation radicals. The results of ensuing experimentation show that steric as well as electronic factors combine to control proton transfer rates in amine cation–stilbene anion radical pairs. The unexpectedly large role played by steric factors in this system is perhaps due to the fact that proton transfer is occurring in highly ordered singlet-contact ion–radical pairs. A summary of the α-substituent effects on tertiary aminium radical kinetic acidities derived from Lewis' efforts is given in Table 3.

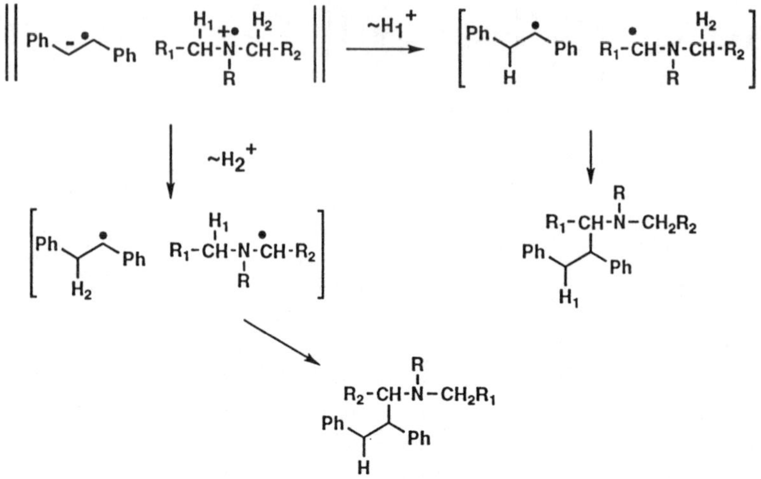

Scheme 15.

The general synthetic utility of styrene– and stilbene–amine SET-promoted photoreactions had not been explored until recent investigations by Lewis and his co-workers.[94,95] As is true with most SET photochemical processes or for that matter any processes involving C-C bonding between radical intermediates, the chemical efficiencies of amine–arylalkene photoreactions increase when they occur in intramolecular systems. However, in contrast to the modestly efficient intermolecular photoaddition reactions observed to occur between tertiary amines and styrenes or stilbenes, linked β-styrylamino alkanes of general structure **65** do not undergo analogous C–C bond-forming processes

Table 3. Relative αC–H Kinetic Acidities of Tertiary Amine Cation Radicals of Y_2NCH_2R Measured by Product Ratios from Amine–Stilbene Photoadditions

R	*Relative Rates per H*
H	1.1
Me	0.5
CO_2Me	2.3
Ph	1.0
$CH{=}CH_2$	0.5
$C{\equiv}CH$	111

when irradiated. The lack of reactivity in these cases does not appear to be due to the inability of the excited reactants to participate in intramolecular SET since the linked systems readily form fluorescent intramolecular exciplexes when irradiated in nonpolar solvents. The lack of reactivity has been attributed to the inability of intermediate zwitterionic diradicals to undergo intramolecular proton transfer (**66** → **67**).

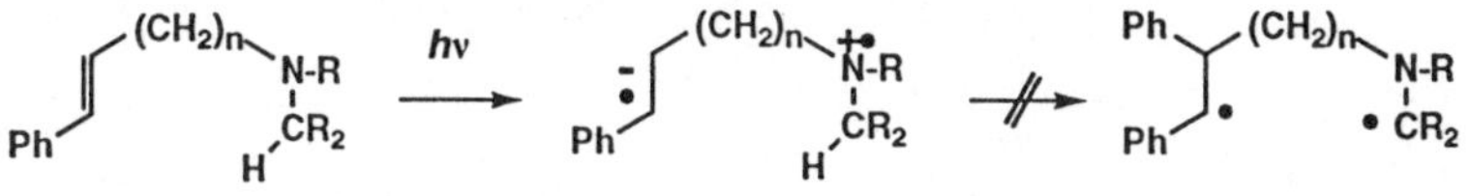

These observations should be contrasted to those made in closely related studies by Aoyama and his co-workers.[96,97] Their efforts have shown that photocyclization reactions do occur in some tertiary α-styrylaminoalkanes. An example of this is found in the pyrrolidine ring–forming cyclization of **68** shown in Scheme 16.

It is interesting that secondary amine analogs **69** of the linked tertiary amines **65** participate in efficient SET-promoted photocyclizations.[94,95] These processes result in formation of products containing both α- and β-phenethylamine grouping (i.e., **70** and **71**). Lewis has shown that the intramolecular additions are highly stereospecific.[98] Like their intermolecular counterparts (see Scheme 13), photoreactions occur by syn addition of the N and H components across the styrene double bond. This is exemplified by the transformations of the *cis*- and *trans*-aminostyrenes **72** and **74** to the respective *trans*- and *cis*-deuteriopiperidines **73** and **75**. Finally, Lewis has nicely demonstrated that the intramolecular process can be used to prepare compounds possessing the benzazepine skeleton (e.g., **76** → **77**) and consequently that the preparative potential of this chemistry is high.[99]

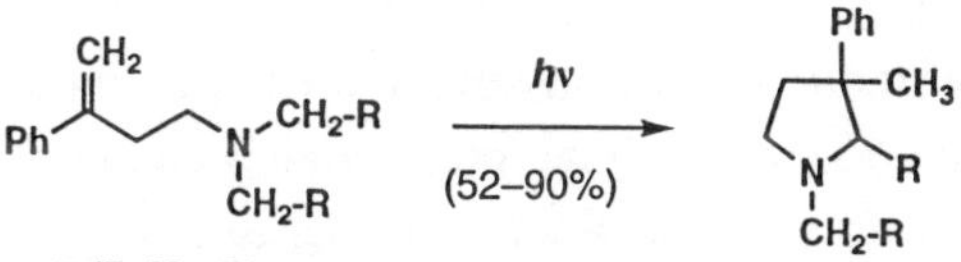

Scheme 16.

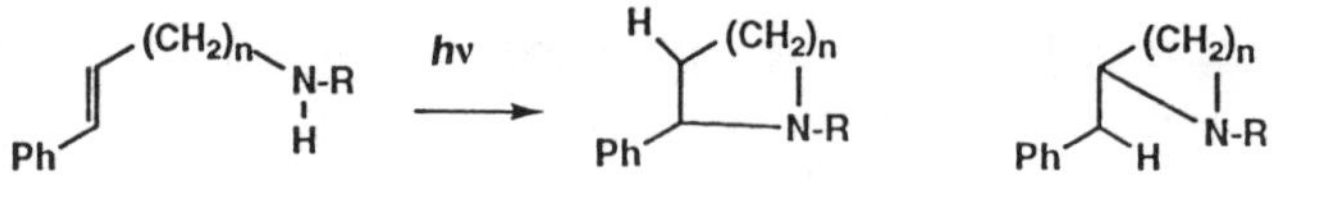

3.4. Amine–Ketone SET Photochemistry

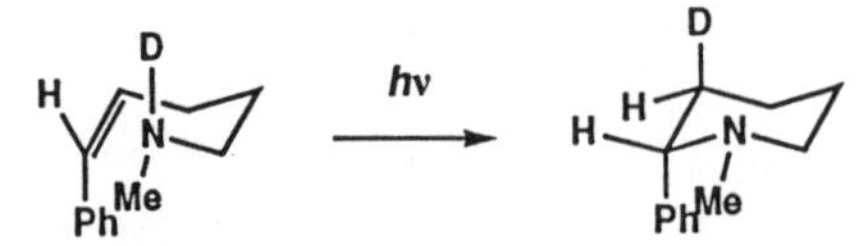

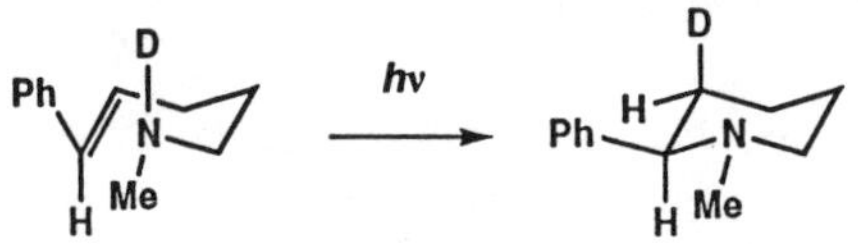

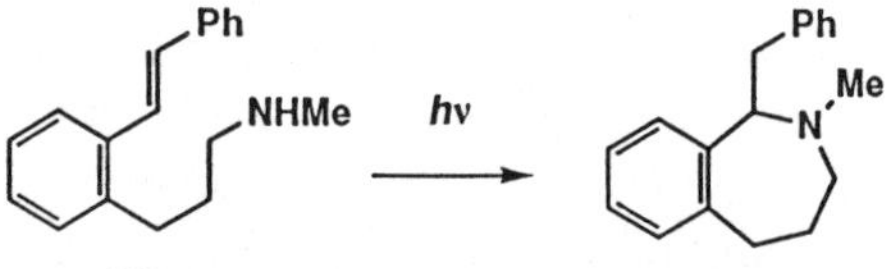

A number of early observations made in the laboratory of Cohen[100,101] convincingly demonstrated that amines participate as electron rather than H-atom donors to singlet and triplet excited states of ketones (Scheme 17). Processes promoted in this fashion and driven by proton transfer in ion–radical pairs lead to the production of the same types of radical intermediates that would have been formed by the more familiar (at that time) H-atom abstraction reactions of ketone excited states. A number of well-designed and compelling experiments were employed to arrive at the view that SET pathways predominate in these systems. Several key observations include the finding that (1) amines readily reduce ketones which have lowest-energy π–π^* states,[102–110] (2) the efficiencies of ketone photoreduction are inversely related to the amine oxidation potentials,[111] and (3) ion–radical intermediates can be detected in these processes by a variety of magnetic resonance and absorption spectroscopic techniques.[112–115] A now dated yet still excellent summary of this early work can be found in a thorough review written by Cohen, Parola, and Parsons.[100]

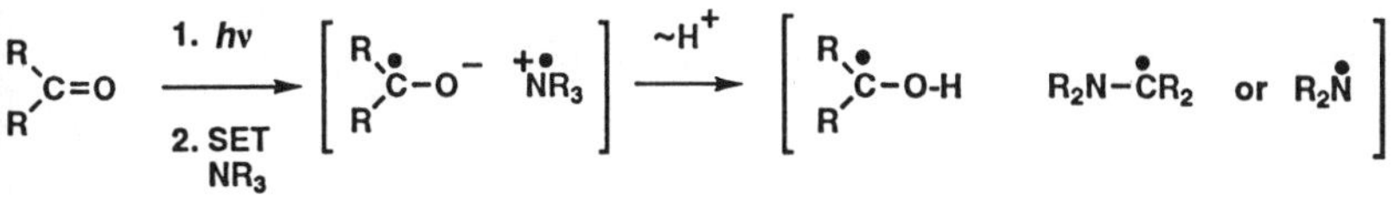

$$\underset{R}{\overset{R}{\diagdown}}C{=}O \xrightarrow[2.\ SET\ NR_3]{1.\ h\nu} \left[R_2\dot{C}{-}O^- \quad \overset{+\bullet}{N}R_3 \right] \xrightarrow{\sim H^+} \left[R_2\dot{C}{-}O{-}H \qquad R_2N{-}\dot{C}R_2 \text{ or } R_2\dot{N} \right]$$

Scheme 17.

The chemical consequences of the sequential SET proton transfer pathways followed in the excited-state chemistry of amine–ketone systems have been uncovered by a number of ensuing efforts in this area. In general, the major reactions observed involve ketone reduction leading to alcohol-containing adducts and amine oxidation producing adducts or imine precursors of carbonyl products. An example is found in the irradiation of benzophenone in solutions containing methylamine, which leads to benzpinacol and the labile imine $H_2C{=}NH_2$.[116] This process is most likely driven by N–H proton transfer from the amine cation radical to the benzophenone anion radical. In contrast, αC-H proton transfer is involved in the photoaddition reaction of benzophenone with Et_3N, which gives the adduct **78**.[117] SET-Induced dealkylation processes have been detected in amine–ketone photochemistry, as seen in the reaction of benzophenone and its derivatives with several tertiary amines.[118] Although several alternative mechanisms can be imagined for these reactions, the route shown in Scheme 18 involving α-amino radical oxidation by ground-state ketone appears to be most reasonable.[101,119,120] Firm support for this proposal is found in the results of laser spectroscopy studies conducted by Das and Bhattacharyya,[121] which show that ketyl radical anion **79** formation occurs in the photoreactions of benzophenone derivatives with Et_3N by both a fast (nanosecond time scale) and a slow (microsecond time scale) process. The slow reaction, which has a rate linearly dependent on ground-state benzophenone concentration and dependent on the reduction potential of the benzophenone derivatives, has been assigned as the reduction of the ketone by the α-amino radical derived from Et_3N.

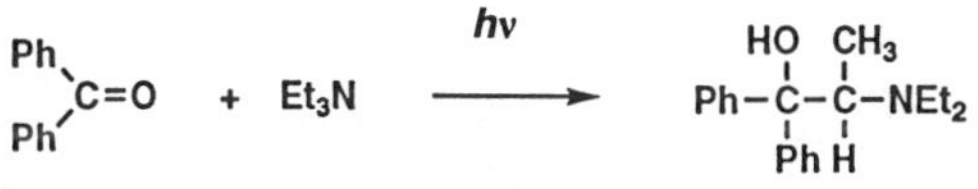

$$Ph_2C{=}O + Et_3N \xrightarrow{h\nu} Ph{-}\underset{Ph}{\overset{HO}{C}}{-}\underset{H}{\overset{CH_3}{C}}{-}NEt_2$$

78

Intramolecular SET in linked aminoketones serves as the activator of photocyclization and photofragmentation reactions which closely re-

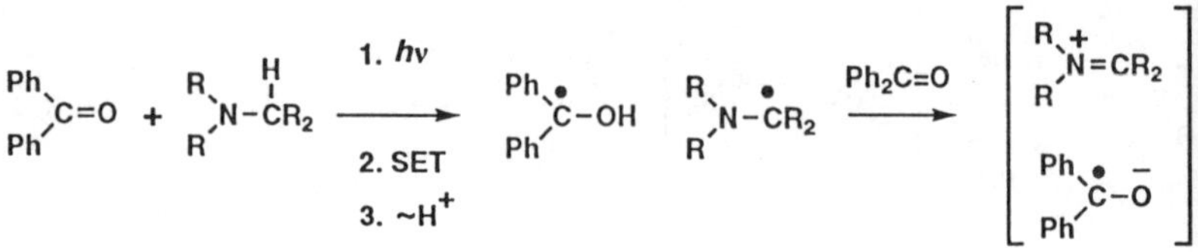

Scheme 18.

semble the familiar Norrish Type-II processes. Examples of reactions of this type leading to small-ring N-heterocycle synthesis[122–124] and fragmentation[125–127] are given in Schemes 19 and 20.

Among the most thorough of all studies probing the intimate details of sequential SET proton transfer sequences in amine–ketone photochemistry are those conducted by Peters and his co-workers. A brief summary of the results of their efforts will be given here and a more thorough discussion is included later in this chapter. By use of picosecond laser flash photolysis techniques, this group has been able to delineate the dynamics of both the SET and the proton-transfer events in these pathways, to identify the nature of ion–radical pairs that serve as intermediates, and finally to determine how orientation of the partners in these ion–radical pairs can govern the regioselectivities of proton transfer. In their initial studies,[128–130] Peters and his collaborators observed that SET from amines to benzophenone triplets occurs at diffusion-controlled

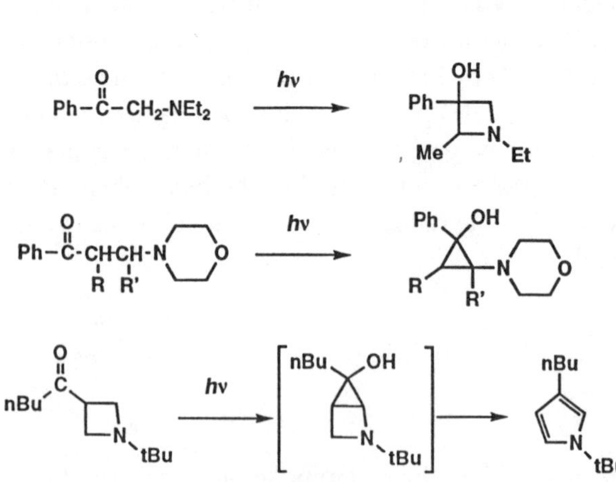

Scheme 19.

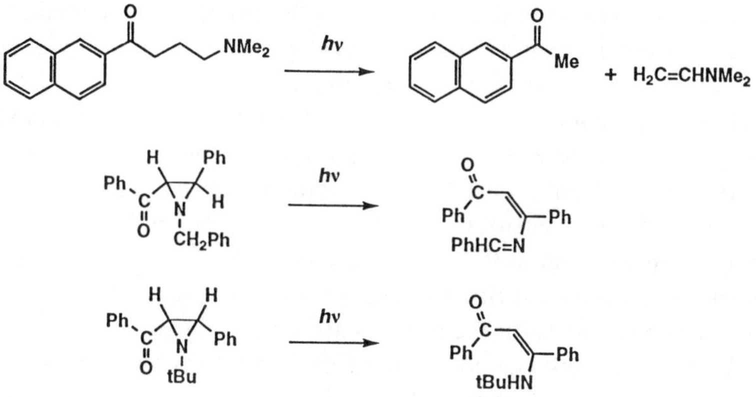

Scheme 20.

rates even in solvents of low polarity [e.g., for dabco $k_{SET} \times 10^{-10} = 1.3$ (C_6H_6), 0.4 ($CHCl_3$), 1.1 (THF), 1.7 (MeCN)]. SET in these systems leads to generation of solvent-separated ion–radical pairs (SSIRP), which then collapse (300 ps) to form contact ion–radical pairs (CIRP) (Scheme 21). Proton transfer in the CIRP occurs at rates that are inversely dependent on solvent polarity [e.g., for Ph_2NMe, $k_{-H^+} \times 10^{-6} = 5$ (C_6H_6) and 1.6 (pyridine)].

The equilibrium between the unprotonated and protonated ketyl anion radical was found to depend on both the nature and the concentration of the amine [e.g., K(MeCN) = 2.7 ($PhNMe_2$) and 0.1 ($PhNEt_2$)]. While this effect was attributed to the influence of high amine concentrations on the

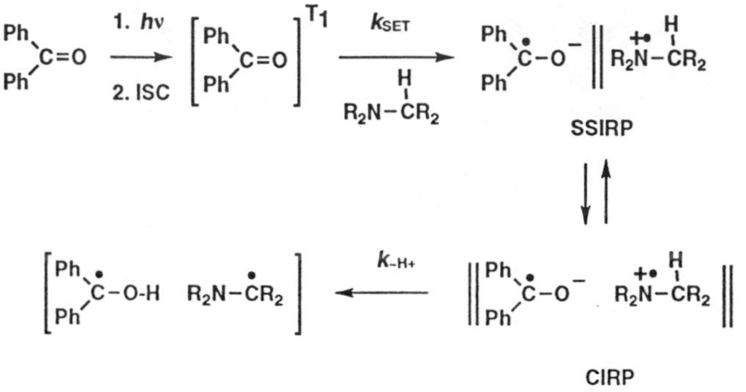

Scheme 21.

solvent dielectric, it is also possible that the variations seen could be a result of the partitioning of the proton between the benzophenone anion radical (pKa = ca. 10)[131,132] and the amine. Polar protic solvents (e.g., EtOH) and oxophilic metal cations (e.g., $LiClO_4$) were also observed to have a pronounced effect on the equilibrium between the SSIRP and CIRP formed from triplet benzophenone and tertiary anilines.[133–135] These results are consistent with the view that H-bonding and metal cation complexation at the oxygen center in the benzophenone anion radical leads to its stabilization, thus removing the need for charge-attractive contact with the amine cation radical partner.

A final noteworthy result arising from the efforts by Peters and his co-workers has come from studies probing the rates of proton transfer between the *N*-methylacridine cation radical **80** and the anion radicals of benzophenone and anthraquinone.[129] Deuterium isotope effects on the rates of proton transfer in the 9,9-d_2 and N–CD_3 analogs were used to demonstrate that deprotonation of **80** by the benzophenone anion radical is selective for the N–CH_3 position while both the 9,9-H_2 and N–CH_3 protons in **80** are abstracted with nearly equal rates by the anthraquinone anion radical (Scheme 22). These observations suggest that proton transfer in CIRPs occurs over short distances and, consequently, that regioselectivities can be governed by ordering of the ion–radical partners in these intimate pairs. According to this reasoning, the CIRP between **80** and the benzophenone anion radical is likely to have a structure represented by **83**, in which the oppositely charged centers are in close

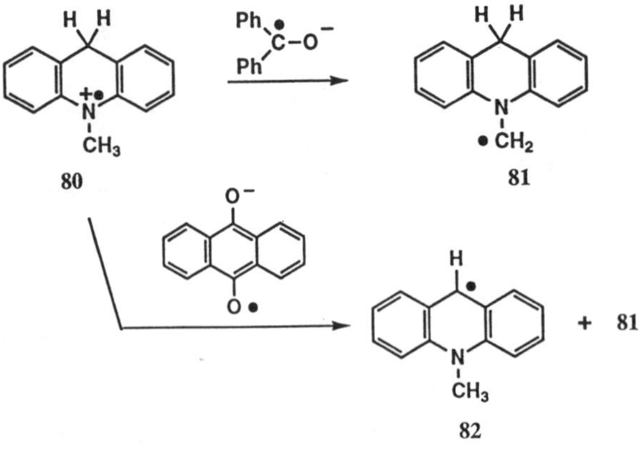

Scheme 22.

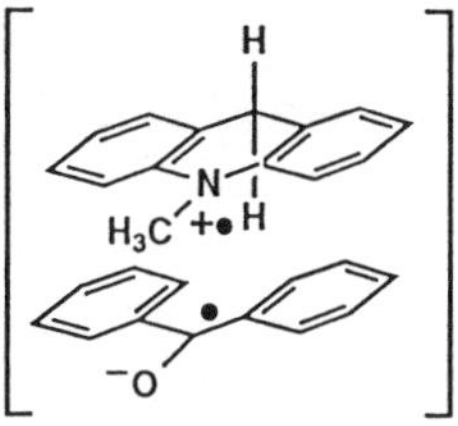

83 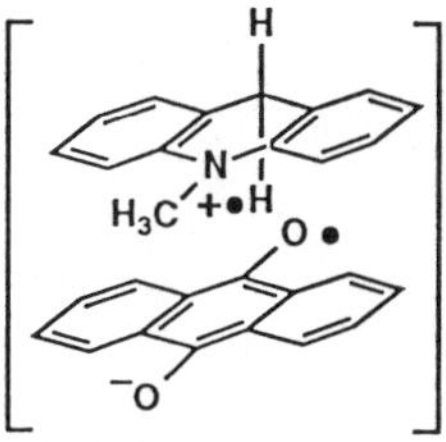 84

contact. As a result, the N–CH_3 protons and basic oxygen center are proximate and proton transfer prefers the N–CH_3 site. In contrast, the anthraquinone anion radical contains a delocalized negative charge, suggesting that proton transfer in the CIRP **84** from both the N–CH_3 and 9,9-H_2 positions of the acridine cation radical are likely. If this observation and its interpretation are both correct and generalizable, effects other than those of steric, stereoelectronic, or electronic origin could well play a role in governing the kinetic acidities of αC-H protons in amine cation radicals as measured by the rates of proton transfer in CIRP intermediates.

4. ALTERNATIVE FRAGMENTATION REACTIONS OF AMINIUM RADICALS

In the preceding section, a number of SET-promoted photoreactions in which amine radical cation intermediates undergo αC–H or N–H deprotonation have been described. The factors governing this type of ion–radical reactivity is related to both thermodynamic and kinetic acidities resulting from the delocalization of positive charge density from nitrogen into the N–H and αC–H bonds and, in the case of αC–H deprotonation, by stabilization of the resulting carbon-centered radical through p_N–p_C orbital overlap. With this view in mind, it is possible to understand the facility of a number of closely related amine cation radical fragmentation reactions in which electrofugal groups other than protons serve as leaving groups. Among the best studied of these are amine cation radical decarboxylations (-H^+, -CO_2), retro-Aldol-like cleavages (-H^+, -$R_2C{=}O$), and desilylations (-SiR_3^+). In each of these processes a base or nucleophile participates to remove the electrofugal group, thus facilitating formation of α-amino radical intermediates.

These reaction types find analogy in the chemistry of arene cation radicals. SET-Induced photodecarboxylations of phenylacetic acid derivatives, for example, are well known reactions. The transformation of

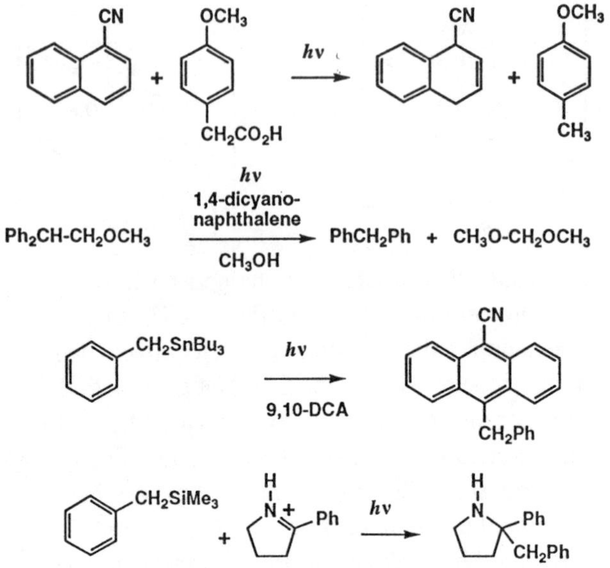

Scheme 23.

p-methoxyphenylacetic acid to the toluene derivative (Scheme 23) sensitized by 1-cyanonaphthalene is representative of this reactivity.[136] Also, C–C bond cleavage facilitated by a β-oxygen center[137] and carbon–metal bond fragmentation in destannation[138] and desilylation[139,140] reactions of arene cation radicals exemplify facile cation radical α-fragmentation processes (Scheme 23).

4.1. Photodecarboxylation of α-Amino Acids

Investigations by Davidson and his co-workers in the 1970s uncovered a number of interesting SET-promoted reactions of α-amino acids and related α-oxy and α-thio analogs driven by cation radical decarboxylations. Stimulated by an earlier controversy over the mechanism for photoinduced decarboxylation reactions of *N*-(2-nitrophenyl)glycines,[141,142] Davidson and his group[143] explored the photochemistry of nitroarenes and *N*-(2-chlorophenyl)glycine as a model system. They found that irradiation of a MeCN solution of a mixture of these substances leads to efficient decarboxylation of the amino acid and production of 2-chloroaniline. An SET mechanism was proposed for this

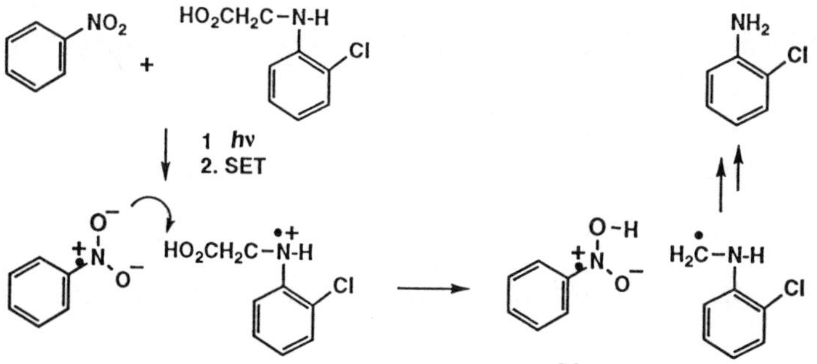

Scheme 24.

transformation in which the glycine radical cation undergoes nitroarene-anion-radical-assisted decarboxylation to form the α-amino radical intermediate **85** (Scheme 24). Although a different route was proposed for termination of this process, it is likely that oxidation of **85** by SET to the nitroarene or intermediate **86** is responsible for ultimate formation of the aniline derivative.

Davidson observed that decarboxylation of *N*-(2-chlorophenyl)glycine can also be SET-photosensitized by benzophenone and quinones.[144,145] For example, irradiation of a benzene solution containing this amino acid and benzophenone, 9,10-anthraquinone, 9,10-phenanthraquinone, or tetrachloro-*p*-benzoquinone leads to evolution of CO_2 and production of both 2-chloroaniline and its *N*-methyl derivative. The N–H to N–CH_3 aniline product ratio can be increased by inclusion of the good radical reducing agent PhSH in the photolysis mixtures. These observations are again consistent with a sequence for decarboxylation that is initiated by SET to the excited ketone acceptors and a termination route involving either α-amino radical reduction by H-atom abstraction or oxidation by SET to the ground-state ketone. In their efforts, Davidson and his collaborators noted that (1) both phenylacetic and non-aryl-substituted α-amino acids are unreactive under the same photolysis conditions where *N*-(2-chlorophenyl)glycine reacts efficiently, and (2) the quantum efficiencies of the decarboxylation reactions for various Ar-X–CH_2CO_2H substrates vary in the series Ar–X = 2-Cl–C_6H_4–NH > C_6H_5S > C_6H_5O, and for various ketone acceptors quantum yields vary in a series that parallels their triplet excited-state reduction potentials. These observations, while offering further support for the SET nature of the processes, also suggest the need for arene-conjugated heteroatom electron-donating moieties. It should be pointed out that the absence of photodecarboxylation reactivity of non-*N*-aryl-substituted α-amino acids might also be a result of their more preferable (i.e., higher pKa) existence in zwitterionic forms which lack nonprotonated nitrogen donor sites.

4.2. Retro-Aldol Cleavages of β-Amino Alcohols

Radical cations arising by one-electron oxidation of β-amino alcohols have been observed to undergo C-C bond fragmentation in a manner analogous to decarboxylations of α-amino acid derivatives. The first observation of this reaction type was uncovered by Davidson and Orton[147] in their investigations of the SET photochemistry of 2-aminoethanols with carbonyl and heterocyclic and arene acceptors. An example is seen in the photochemistry of anthraquinone with the amino alcohols **87** and **88**, where benzaldehyde and the anilines **89** and **90** are produced on irradiation of benzene solutions (Scheme 25). In these

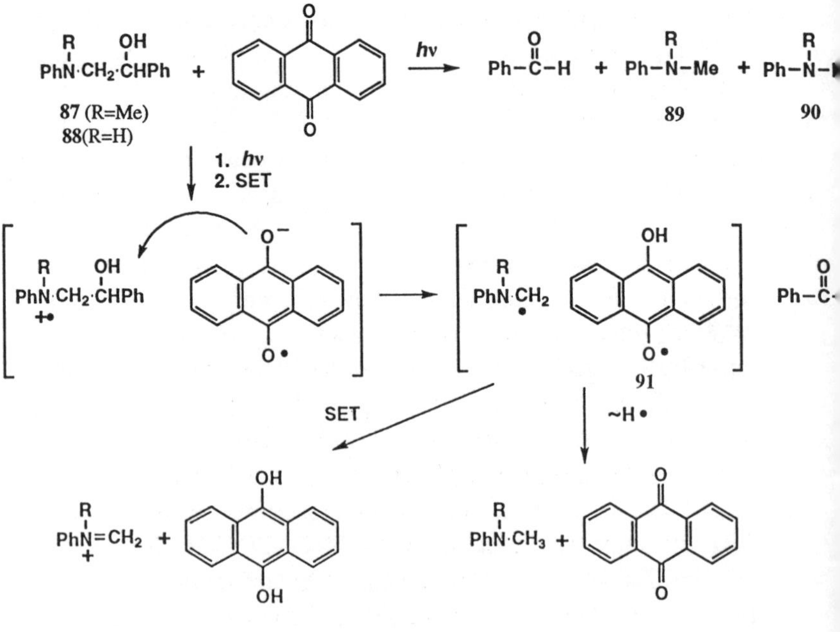

R OH
PhN·CH₂·CHPh +
87 (R=Me)
88(R=H)
hv
Ph–C(=O)–H + Ph–N(R)–Me + Ph–N(R)–
89 90
1. hv
2. SET
PhN·CH₂·CHPh (R, OH, +•)
PhN·CH₂ (R, •)
91
SET
~H•
PhN=CH₂ (R, +) +
PhN·CH₃ (R) +

Scheme 25.

processes, the quinone is consumed to a lesser extent than are the amino alcohols, suggesting that the ketone serves in part as a sensitizer for the fragmentation reaction. This is consistent with a mechanistic sequence for formation of the respective tertiary and secondary amines **89** from **87** and **88**, one in which H-atom transfer occurs between the α-amino and hydroquinone radical **91** intermediates in competition with α-amino radical oxidation.

A more recent and intense study of β-amino alcohol SET photofragmentations has been conducted by Whitten and his co-workers.[17,148–154] In efforts (1) using a number of different amino alcohols, SET-acceptors, solvents, and (2) probing reaction quantum efficiencies, and the nature and lifetimes of intermediates, a detailed picture of the mechanistic pathways followed in these processes has emerged. In general, Whitten has demonstrated that the β-amino alcohols, such as the diastereomeric morpholine derivatives **92** and **93**, undergo clean (ca. 90%) C-C bond cleavage reactions promoted by irradiation of a variety of SET acceptors including thioindigo, 9,10-dicyanoanthracene, and 2,6,9,10-tetracyanoanthracene. The overall process in each case involves reduction of the acceptor and fragmentation to produce carbonyl and amine products. The quantum efficiencies of the processes are generally low (10^{-1}–10^{-4}) owing to the competition between electron backtransfer and fragmentation in the singlet ion–radical pairs **94** (Scheme 26). From observed solvent, acceptor, and amino-alcohol stereochemical effects, Whitten has concluded that the highest photofragmentation quantum efficiencies are obtained when (1) the solvent allows for contact ion–radical pair formation, (2) the acceptor anion radical is basic, and (3) steric interactions between amino alcohol cation radical substituents in the fragmentation transition state are minimized. These results suggest that the fragmentation process occurs by a route in which acceptor anion radical deprotonation and C–C bond cleavage are concerted and where antiorientation of the hydroxyl and aminium nitrogen functions is preferred.

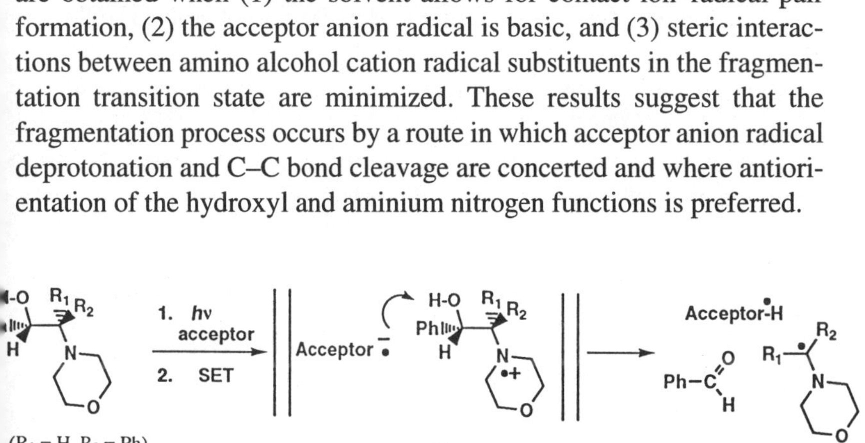

1. hν acceptor
2. SET

(R_1 = H, R_2 = Ph)
(R_1 = Ph, R_2 = H)

94

Scheme 26.

4.3. α-Silylamine Cation Radical Fragmentations

Nucleophile-assisted α-silylamine cation radical α-fragmentation is yet another type of reaction available to nitrogen-centered charged-radical intermediates. Prior to the efforts of Mariano, Yoon, and their co-workers (described in detail in the next section), little was known about processes of this type, which lead to the generation of α-amino radicals. However, more information was available on the mechanistic features of related allylsilane and benzylsilane cation radical reactions.[155–159] In earlier investigations, this group had shown, in studies of SET-promoted photoaddition and photocyclization reactions of 1-pyrrolinium salts, that both allyl- and benzylsilane cation radical intermediates undergo exclusive desilylation by transfer of a trialkylsilyl group to either MeOH or MeCN as nucleophiles. These pathways are followed exclusively even though competitive deprotonation reactions of the intermediates are possible. Information about the relative rates of benzylsilane cation radical deprotonation, decarboxylation, and desilylation has come from more recent collaborative work by the groups headed by Albini and Mariano[160] in which product distributions of photoaddition reactions between 1,4-dicyanonaphthalene and a variety of *p*-xylene derivatives were determined (Scheme 27). The order of cation radical α-fragmentation reactivity determined in this work is $-SiMe_3 > -CO_2H > -H$.

The first example of a SET-promoted photochemical reaction involving a tertiary α-silylamine is found in the addition of $Et_2NCH_2SiMe_3$ to 9,10-dicyanoanthracene.[84] This process leads to production of the non-silicon-containing adduct **52** (see Section 3.2.2), the result of selective transfer of the TMS group to solvent MeCN from the silylamine cation radical intermediate.[158,159] Based on this result, a preliminary conclusion could be made that desilylation is the preferred reaction pathway avail-

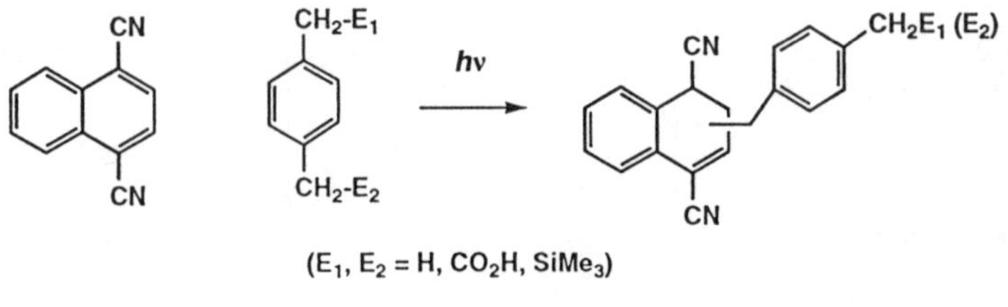

CN CN CH_2-E_1 CH_2-E_2 hν CN CN CH_2E_1 (E_2)

(E_1, E_2 = H, CO_2H, $SiMe_3$)

Scheme 27.

able to α-silylamine cation radicals. Thus, the thermodynamic stabilization offered in these intermediates by p_N-$\sigma_{C\text{-}Si}$ orbital overlap, which leads to a lowering in the oxidation potentials of the amines,[22–26] appears to result in kinetic instability. However, as will be seen below, this conclusion should be viewed with caution since the chemoselectivity of tertiary α-silylamine cation radical reactions (i.e., desilylation versus deprotonation) is intimately related to the nature of silophiles and bases present under the SET photochemical reaction conditions employed.

5. SET-PROMOTED PHOTOREACTIONS BETWEEN TERTIARY AMINE AND α,β-UNSATURATED CARBONYL COMPOUNDS

5.1. Early Exploratory and Mechanistic Studies

As with their saturated counterparts, excited states of α,β-unsaturated carbonyl compounds (ketones and esters) participate as acceptors in SET processes with tertiary amine donors. The first examples of transformations activated in this fashion are found in exploratory studies in the 1960s by Cookson and his co-workers.[161] The reactions discovered at that time (Scheme 28) clearly demonstrate that Et_3N photoadds to a variety of α,β-unsaturated ketones and esters to produce adducts resulting from bonding at the β-carbon of the carbonyl substrate and the amine

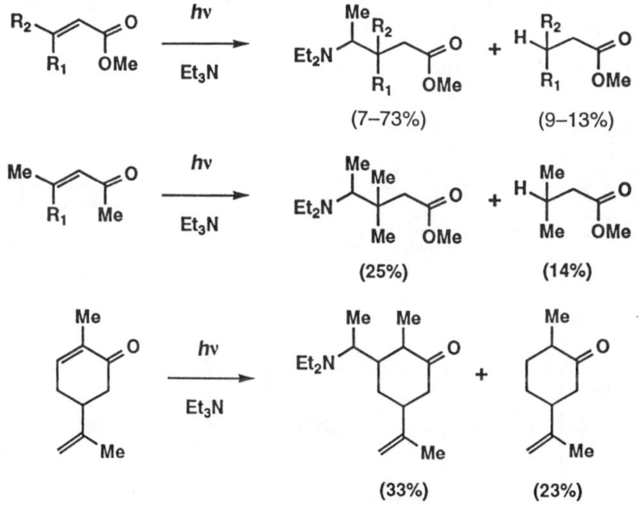

R2 R1 OMe O; hν, Et3N → Et2N Me R2 R1 OMe O (7–73%) + H R2 R1 OMe O (9–13%)

Me R1 Me O; hν, Et3N → Et2N Me Me Me OMe O (25%) + H Me Me OMe O (14%)

Me O Me; hν, Et3N → Et2N Me Me O Me (33%) + Me O Me (23%)

Scheme 28.

α-carbon in low to moderate yields. In each case the photoaddition process is accompanied by photoreduction to produce the saturated ketone or ester.

These processes, promoted by irradiation of the carbonyl compounds in neat Et_3N, obviously have an interesting structural outcome even though the efficiencies are low. Despite this feature, amine-unsaturated ketone photoaddition did not receive further study until the 1980s, when Pienta and his collaborators[162–164] made a number of key observations relating to the mechanistic route followed. These workers found, for example, that irradiation of a MeCN solution of cyclohexenone **95** in the presence of Et_3N leads to production of the β-adduct **96** along with the reduction product **97** and enone dimer **98**. The yields of **96** and **97** were found to increase and those of **98** decrease with increasing Et_3N concentrations. This result is consistent with interaction of Et_3N with an excited-state intermediate common to both the dimerization and photoaddition and photoreduction pathways. In addition, Pienta noted that naphthalene, a known quencher of cyclohexenone triplets, inhibits formation of all products equally. Finally, CIDNP studies of photoprocesses occurring between cyclohexenone **95** and both dabco and Et_3N showed that the initial event involving the amine and cyclohexenone triplet is SET rather than H-atom transfer. These data led Pienta to propose that the enone–amine photoreactions are initiated by SET to the triplet ketone excimer (established more conclusively to be the monomeric triplet in later efforts by Schuster and his collaborators[33,165–167]) and most probably involve the ensuing sequence of proton transfer and radical coupling (Scheme 29).

No information was available in these early efforts to define in more detail the steps in the processes following the SET-event. Thus, while it

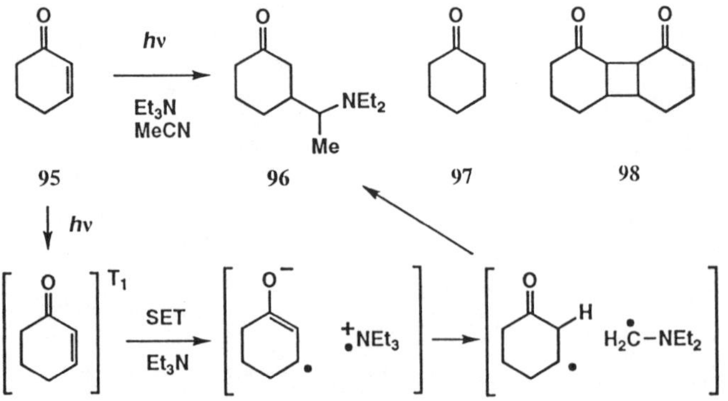

Scheme 29.

was clear that amine cation radical deprotonation, enone anion radical protonation, and C-C bond formation are all required for adduct formation, the timing of these steps was unclear. Also, information about synthetically related issues, such as the regiochemistry of reactions with unsymmetrically substituted amines, the control of reaction efficiency in inter and intramolecular systems, and the stereochemistry of the process when chiral centers are formed remained to be explored. Finally, no information about the sequences involved in the photoreduction reactions came from these early mechanistic studies.

5.2. α-Silylamine–Enone Photoaddition and Photocyclization Processes

5.2.1. Competition between α-Silylamine Cation Radical Desilylation and Deprotonation

Interest in mechanistic and synthetic features of the amine–enone SET-promoted photoaddition reactions uncovered by Cookson and preliminarily studied by Pienta and Schuster has led to a number of recent investigations by Mariano, Yoon, and their co-workers These groups initially explored the photoreactions of the tertiary α-silylamine, $Et_2NCH_2SiMe_3$, with a variety of cyclic conjugated enones. The aim of the work was to determine if, as expected, rapid desilylation of the intermediate α-silylamine cation radical **99** would convey high degrees of regiocontrol to the addition reactions. The results of exploratory investigations[168] indicated that this expectation was only partially upheld. As seen for the prototypical process shown in Scheme 30, two types of adducts are produced in these photoadditions. Specifically, the TMS- and non-TMS adducts, **102** and **101**, are both generated in ratios which are altered by deceptively subtle changes in solvent.

Me–HĊ(Et)–N–CH(H_a)–$SiMe_3$
$-H_b^+$
$-SiMe_3^+$
Me–HC(H_b)(Et)–$\dot{N}^{+}$–CH(H_a)–$SiMe_3$
Me–HC(H_b)(Et)–N–ĊH(H_a)
Me–HC(H_b)(Et)–N–ĊH–$SiMe_3$
$-H_a^+$
99

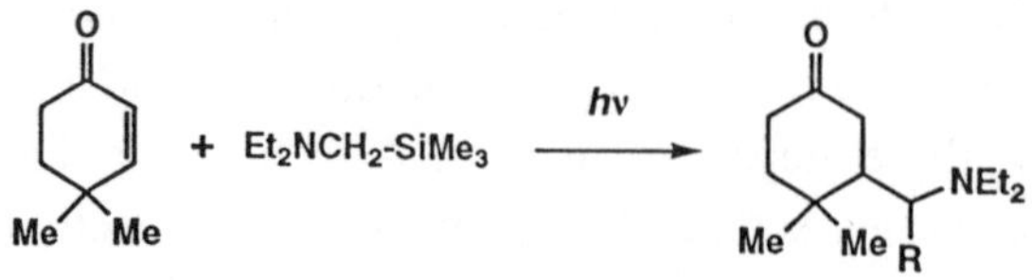

100 + Et_2NCH_2-$SiMe_3$ $\xrightarrow{h\nu}$

	MeCN	MeOH
101 (R=H)	8%	60%
102 ($R=SiMe_3$)	70%	30%

Scheme 30.

The sources of these and other effects on the chemoselectivity (i.e., TMS versus non-TMS adduct formation) of SET photoadditions of $Et_2NCH_2SiMe_3$ was revealed in a thorough study[169] probing the influence of solvents, added salts, and bases on the ratio of adducts arising from reaction with cyclohexenone **100**. For example, the TMS adduct **102** was found to form predominantly in reactions run in low-polarity aprotic solvents (e.g., **101**:**102** = 0.05 in cyclohexane and 0.16 in MeCN) while in more polar, protic media the non-TMS adduct **101** forms preferentially (e.g., **101**:**102** = 2.97 in MeOH and 4.08 in 25% H_2O–MeOH). Also, the presence of high concentrations of the oxophilic cation–containing $LiClO_4$ results in predominant production of the non-TMS adduct (e.g., **101**:**102** = 2.68 in 0.4 M $LiClO_4$ in MeCN). Thus, these results show that the silylamine–enone photoadditions are mechanistically more complicated than first envisioned and that the medium plays a major role in governing the relative rates of desilylation and deprotonation of the intermediate silylaminium radical **99**.

The observations made in these efforts were interpreted as implicating mechanisms for the photoaddition reaction between $Et_2NCH_2SiMe_3$ and cyclohexenones in which amine cation radical **99** undergoes competitive desilylation and αC-H deprotonation with relative rates which are governed by the solvent-controlled basicity of the enone anion radicals **103** (Scheme 31) and silophilicity of the solvent. Accordingly, proton transfer from **99** to **103** in a contact ion–radical pair to produce the α-silylamino radical **104** is favored in aprotic, nonsilophilic media (e.g., MeCN) where the enone anion radicals should be strongly basic and where the process results in charge annihilation. In contrast, preferential generation of non-TMS radical **105** in protic, highly silophilic solvents (e.g., MeOH, H_2O) or when the oxophilic metal cation Li^+ is present is

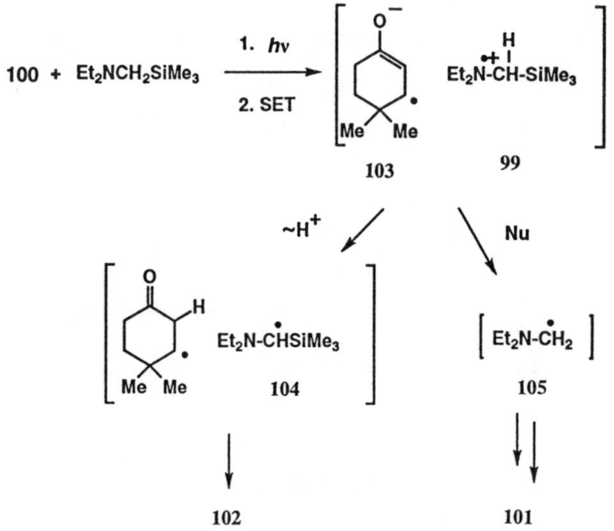

Scheme 31.

a consequence of the attenuated basicity of the enone anion radical **103** because of solvation (H-bonding) or coordination effects.

That acceptor anion radical base strength is an important factor in governing the relative efficiencies of silylamine cation radical deprotonation versus desilylation is further evidenced by the observation that the **101**:**102** ratio from photoreaction of **100** and $Et_2NCH_2SiMe_3$ conducted in MeOH decreases with increases in the amine concentration (e.g., 6.51 at [amine] = 0.04 M and 1.94 at [amine] = 0.65 M). Moreover, the exclusive formation of non-TMS adducts from SET-promoted photoadditions of $Et_2NCH_2SiMe_3$ to the acceptors 9,10-dicyanoanthracene[84] (see above), acenaphthenoquinone,[170] and *N*-methylphthalimide[171] in MeCN is consistent with the proposal that counteranion radical base strength plays a key role in determining the chemoselectivity of α-silylamine cation ion–radical reactions. Accordingly, unlike enone anion radicals whose conjugated acids (1-hydroxyallyl radicals) have approximate pKa's (H_2O) of 10,[172,173] those of DCA (pKa < 0),[79,86] α-diketones (pKa (H_2O) = ca. 5)[172] and phthalimides (pKa << 7)[173] are more acidic, demonstrating the lowered basicity of the corresponding anion radicals.

The ability to control the nature and regiochemistry of silylamine cation radical fragmentation reactions by the choice of solvent has served as a key component in the design of new and highly efficient SET-

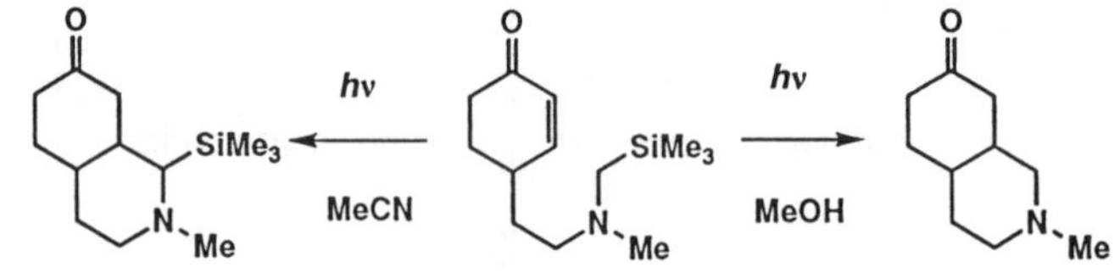

Scheme 32.

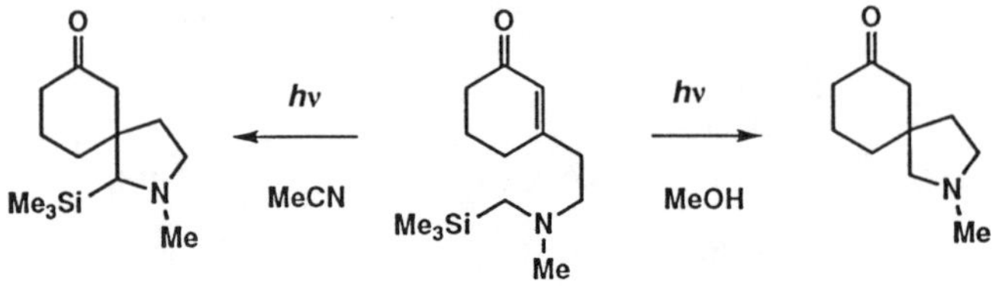

promoted photocyclization reactions. This is demonstrated by observations made in studies of the phototransformations of the silylaminoethylcyclohexenones **106** and **107** (Scheme 32), where cyclizations occur to give non-TMS products cleanly on irradiation in MeOH while the TMS-containing spirocyclic products are formed exclusively when MeCN is used as solvent.[174,175] The much higher degrees of chemoselectivity in these processes as compared to their intermolecular counterparts is presumably due to the fact that low (ca. 1×10^{-3} M) aminoenone concentrations are used in the intramolecular systems. Consequently, amine-induced deprotonation of the intermediate amine cation radical is less favorable.

5.2.2. Probing αC–H Kinetic Acidities of Amine Cation Radicals

During the course of investigations devised to explore more fully the scope of the silylamine SET photoaddition and photocyclization reactions, an interesting control of carbon–carbon bond formation regiochemistry was noted. This is exemplified in the photoadditions of $Et_2NCH_2SiMe_3$ to cyclohexenone **100** and photocyclizations of the silylaminocyclohexenones **106** and **107** in MeCN, where bond formation occurs selectivity at the TMS-substituted amine α-centers. These preferences reflect the relative rates of amine cation radical αC–H deprotonation and their control by substituents.

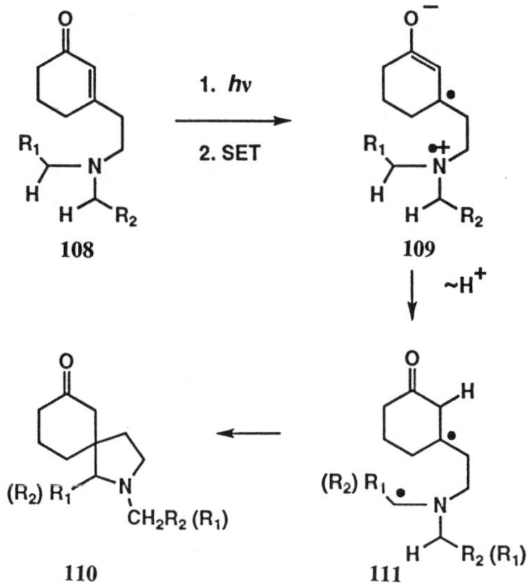

Scheme 33.

If steric factors alone were responsible for governing αC-H kinetic acidities in these systems (see Lewis's results above), it is not obvious why proton loss from the TMS-substituted α-carbon in the aminium radical intermediates would be faster than that at an unsubstituted methyl center. An examination of this issue was undertaken[175,176] by an investigation of the photochemistry of a variety of β-aminoethylcyclohexenones **108** (Scheme 33) where cyclizations occur to produce distributions of regioisomeric products **110** in ratios that reflect the relative rates of proton transfer between the aminium and enone anion radical centers in the intermediate zwitterionic diradicals **109**. Since the transition states for the competitive proton transfers in these reactions are similarly structured, relative kinetic acidities determined by use of this system are not complicated by the issue of orientation of partners in ion–radical pairs. Additionally, product yields in these cyclization reactions are generally high, owing to the absence of cage-escape pathways, which in intermolecular systems could compete with radical coupling.

The data obtained from this study (Table 4) show that substituents play a characteristic role in governing the relative rates of intramolecular proton transfer in the zwitterionic diradicals **109**. Alkyl substituents enhance the proton-transfer rates slightly, the TMS-substituent does so to a greater extent, and conjugating substituents like Ph have the greatest

Table 4. Substituent Control of Kinetic Acidities of Cation Radicals Derived from Tertiary, Y_2NCH_2R Determined by Use of the β-Aminoethylcyclohexenone Photochemical System

	Relative Kinetic Acidity	
R	*MeCN*	*MeOH*
H	0.01	0.01
CH_3	0.02	0.02
$Si(CH_3)_3$	0.1	—
CO_2CH_3	0.5	0.6
C≡C–H	3.9	2.0
Ph	1.0	1.0
CH=CH_2	1.9	3.0

effect. These results provide compelling evidence that electronic as well as steric effects (see Table 3) can significantly contribute to αC-H kinetic acidities of tertiary aminium radicals. Stated simply, for amine cation radical αC-H deprotonation reactions leading to α-amino radicals, radical stabilizing substituents can enhance reaction rates in the same way as they influence thermodynamic acidities (pKa) of these radical cations.[40]

5.2.3. Radical Coupling and Radical Addition Mechanisms

What is unclear from viewing the mechanistic sequence shown in Scheme 31 for the silylamine–cyclohexenone photoaddition reactions is how the processes are terminated. Specifically, carbon–carbon bond formation in these additions in theory can occur by radical pair coupling when the α-amino radical intermediate is generated in a solvent–caged radical pair by proton transfer in a contact ion–radical pair to the enone anion radical. Another alternative involves α-amino radical addition to the unsaturated ketone, a process that should be favorable when the pathway involves proton transfer from the amine cation radical to solvent or amine base. In this event, the α-amino radical should be generated as a free species. In the latter case, the probability of α-amino radical coupling with enone anion radicals would be low as compared to that of their addition to ground state enones,[49,50,177] which are present in relatively high concentrations.

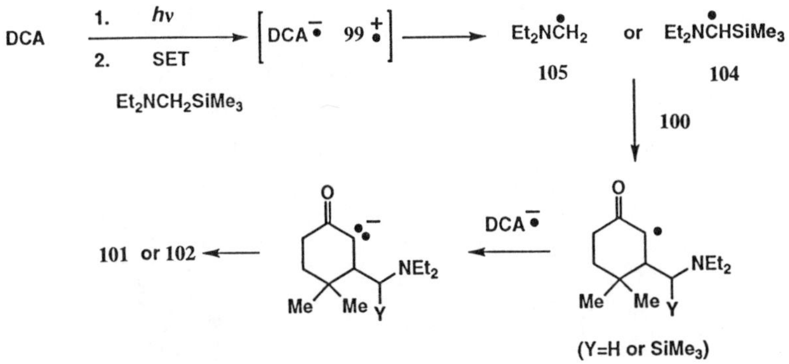

DCA $\xrightarrow[\text{2. SET}]{\text{1. } h\nu}$ [DCA$^{\overline{\bullet}}$ 99$^{\stackrel{+}{\bullet}}$] ⟶ $Et_2N\dot{C}H_2$ (105) or $Et_2N\dot{C}HSiMe_3$ (104)

$Et_2NCH_2SiMe_3$

100

DCA$^{\overline{\bullet}}$

101 or 102

NEt_2

Me Me Y

(Y=H or $SiMe_3$)

Scheme 34.

Observations made in studies by Mariano, Yoon, and their co-workers lend credence to these mechanistic proposals. For example, irradiation of DCA in an MeCN solution containing $Et_2NCH_2SiMe_3$ and cyclohexenone **100** was found to lead to production of adducts **102** and **101** with the non-TMS adduct **101** becoming more predominant as the concentration of **100** is decreased.[169] A reasonable mechanism for this DCA-photosensitized addition, occurring competitively with the direct irradiation process, begins with SET from the amine to DCA^{S_1} and is followed by desilylation and subsequent addition of the resulting α-amino radicals to the enone **100** (Scheme 34). Later studies[178] have demonstrated that termination of these processes occurs through SET from the DCA anion radical to the α-keto radical intermediates, followed by enolate anion protonation.

Firm support for the proposed participation of radical addition pathways in the amine–enone reactions promoted by direct irradiation of the enone has derived from studies[179] designed to probe the photoadditions of *N,N*-dimethylaniline **112** and its α-silyl analog **113** to cyclohexenone **100**. The conceptual basis for this effort revolved about the thought that α-amino radical conjugate addition routes are uniquely characterized by their generation of α-keto radical intermediates. Unlike the enolate anions or enols produced in radical pair coupling mechanisms, α-keto radicals might be trapped by radical addition to appended, electron-rich π-moieties. This possibility exists in the α-keto radical **114**, the intermediate in radical addition routes for aniline **112** and **113** photoreactions with cyclohexenone **100** (Scheme 35). The observed formation by this trapping process of the tricyclic amino-ketone **116** along with the "nor-

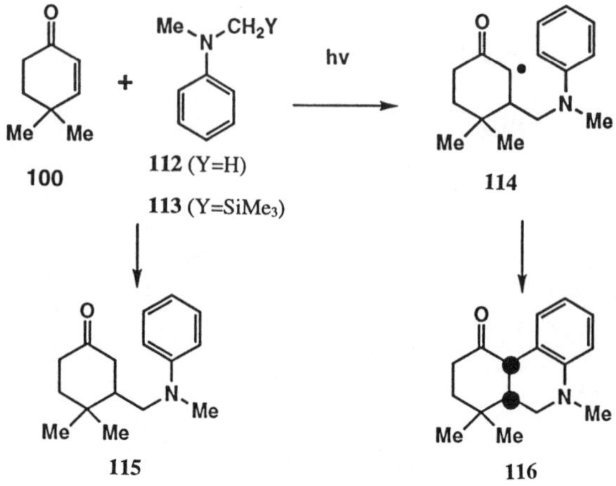

Scheme 35.

mal" adduct **115** in photoreaction of **113** with **100** in MeOH, H_2O–MeCN, and $LiClO_4$–MeCN solutions and nearly exclusively in the DCA-sensitized addition of **113** to **100** signals the existence of radical addition mechanisms in amine and silylamine–enone SET photoreactions.

It is clear from these results that the SET-photosensitization methodology, employing an indirect method for α-amino radical generation and a conjugate radical addition mechanistic route, should be useful in promoting photoreactions in silylamine–enone and –ester systems. It should find special application in initiating cyclization reactions where the intramolecular radical pathways are entropically favorable. Moreover, SET sensitization could be used advantageously in systems where the direct irradiation method is troublesome. For example, since appropriate SET photosensitizers with long wavelength ($\lambda > 320$ nm) absorption properties are available, use of this methodology would avoid formation of enone excited states, which in certain systems (e.g., acyclic enones, dienones) are highly photolabile. Also, the ability to select sensitizers with high excited-state reduction potentials and whose anion radicals are weakly basic extends the range of α-silyl donors that can be used in these reactions and ensures that site-selective desilylation of intermediate amine cation radicals will occur. For example, α-silylamides whose oxidation potentials are too high to readily participate in SET with triplet enones can be oxidized by thermodynamically favored

SET to singlet states of a variety of cyanoarenes (e.g., DCA^{S_1}). Finally, the route followed in SET-sensitized intramolecular reactions, unlike that of their direct irradiation counterparts, does not involve the intermediacy of diradicals, which in some instances can undergo efficient fragmentation.

Illustrative examples of the types of silylamino–enone photocyclization reactions that can be performed by using the SET sensitization method are found in efforts by Mariano, Yoon, and their co-workers[174,175] as seen in the transformations given in Scheme 36. Typically, the conditions used to promote these and related reactions employ low concentrations (ca. 1×10^{-4} M) of DCA as the sensitizer and deoxygenated MeCN–MeOH mixtures as solvent. For example, irradiation ($\lambda >$ 320 nm) of DCA in an MeCN–MeOH solution containing the γ-silylaminoethylcyclohexenone **106** results in efficient production of the diastereomeric hydroisoquinolones **117** and **118** (ca. 6:1) along with a trace (2%) quantity of the hydroisoindolone **119**. A comparison of the stereoselectivity (6:1 *cis:trans*) of the DCA-sensitized photocyclization of **106** with that observed for the reaction promoted by direct irradiation

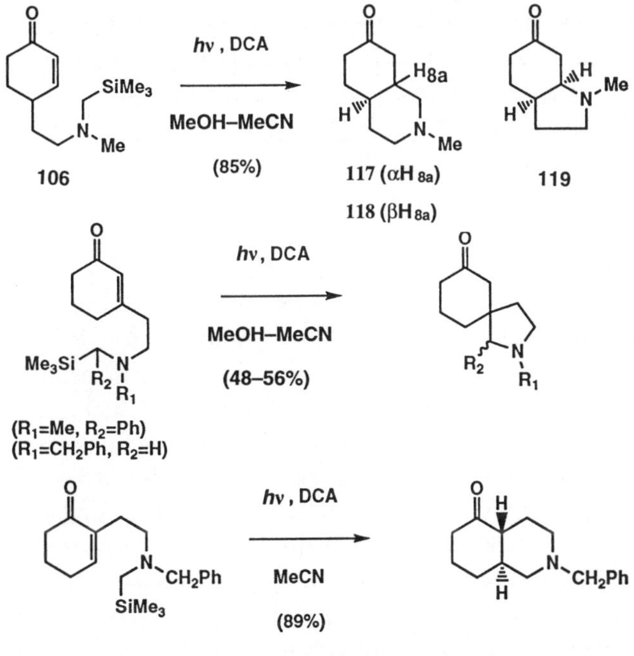

hν, DCA; MeOH–MeCN (85%)

106 → 117 (αH 8a), 118 (βH 8a), 119

hν, DCA; MeOH–MeCN (48–56%)

(R_1=Me, R_2=Ph)
(R_1=CH_2Ph, R_2=H)

hν, DCA; MeCN (89%)

Scheme 36.

of the enone in MeOH (ca. 1:1 *cis:trans*) is informative in that it demonstrates the greater degree of stereoelectronic control associated with intramolecular axial β-additions of radicals to cyclohexenones compared to that of diradical coupling processes.

5.2.4. Competition between α-Amino Radical Cyclization and Oxidation

Information about the scope, limitations, and advantages of the SET-photosensitization method for silylamine–enone cyclization has come from studies conducted during the past several years. One of these efforts focused on the photochemistry of a series of aminoketones and -esters having the general structure **120**.[178] The effects of (1) incorporating unsaturated ketone and ester functions into acyclic frameworks, (2) tether length, (3) carbonyl type, and (4) nitrogen substituents on photocyclization reaction efficiencies were evaluated. The first issue is whether or not in direct irradiation processes intramolecular SET from the amine grouping to the excited enone or unsaturated ester moieties would be competitive with the typically rapid excited-state deconjugation and *cis–trans* isomerization pathways available to these chromophores when found in acyclic environments. As anticipated, direct irradiation of the silylamino ketones **121** (R = CH_3 or CH_2Ph) results only in efficient *cis–trans* isomerization. In contrast, DCA-sensitized photoreactions of these substances in MeCN–MeOH occurs to generate the substituted piperidines **123** in high yields. Similarly, while the related esters **122** (R = CH_3 or CH_2Ph) remain unreactive when directly irradiated, they undergo smooth DCA-sensitized photocyclization to form the piperidine esters **124** along with the pyrrolidine ester **125**.

It is clear that competitive formation of pyrrolidine products **125** in the DCA-sensitized reactions of the silylamino esters is a potentially undesirable feature of this method. Consequently, an investigation of the mechanism for the desilylmethylation processes which lead to these substances was undertaken. Exploratory studies demonstrated that the pyrrolidine:piperidine ratio (**125**:**124**) in the DCA-sensitized photocyclization of **122** increases monotonically with increases in the DCA concentration and that pyrrolidine **125** predominates when the reaction is conducted with oxygenated solutions. Thus, a likely sequence for desilylmethylation involves oxidation of the intermediate α-amino radicals **126** ($E_{1/2}(+)$ = ca. −1 V)[48] by ground-state DCA ($E_{1/2}(-)$ = 0.89 V)[180] or oxygen. Hydrolysis of the resulting formaldiminium cation **127** gives the

120

$h\nu$

121 (Y=Me)

122 (Y=OMe)

$h\nu$, DCA

MeCN–MeOH

(78–90%)

123

124

125 (Y=OMe)

126

127

secondary amine precursor of the pyrrolidine products. The diminished pyrrolidine yields in photoreactions of the unsaturated ketones as compared to those of esters is probably a result of the faster rates of electron-rich α-amino radical additions to ketone- as compared with ester-substituted olefins.

The pathway leading to desilylmethylation is analogous to the familiar ECE sequences followed in two-electron electrochemical oxidations of tertiary amines. Indeed, anodic as well as metal cation [e.g., K_3FeCN_6, $Hg(OAc)_2$, $Pb(OAc)_4$, $Ce(NH_4)_2(NO_3)_6$] oxidations of the silylamino ketones **121** and ester **122** in MeCN provide the respective pyrrolidines exclusively.[181] Parenthetically, Mariano and Zhang[182,183] have shown that this oxidative procedure represents a useful method for generation of formaldiminium cations and, therefore, for promoting well-documented iminium ion cyclization processes (Scheme 37).

The contrast between the DCA-sensitized and electrochemical or metal cation oxidation reactions of the silylamino ketones and esters illustrates one of the advantageous features of the SET photosensitization method for initiating these α-amino radical cyclizations. As with other

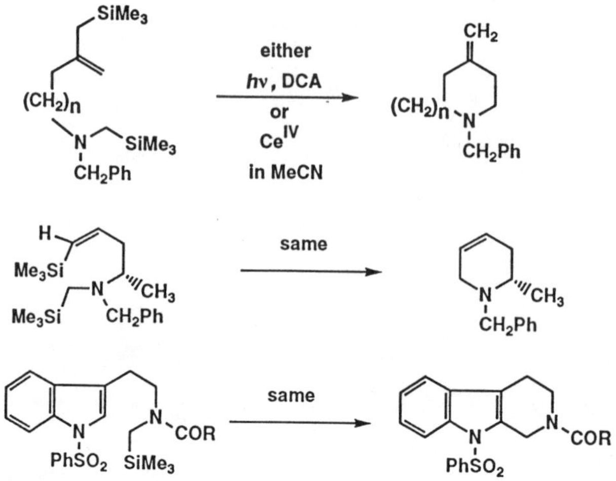

Scheme 37.

photochemical reactions, the conditions used are mild and often compatible with a range of functionality. Of greater importance is the fact that the oxidant in the photochemical procedure is the sensitizer excited state which, owing to the low light fluxes employed, is present in extremely low concentrations. As a result of this, the oxidant has a low probability of promoting secondary oxidation of the transient and easily reduced α-amino radical. The alternate oxidation methods unfortunately bring about uncontrollable two-electron oxidations.

Oxidation of the α-amino radical intermediates by the ground-state sensitizer could be one of the major drawbacks of the SET-photosensitized cyclization reactions. A search to find ways of removing this limitation has uncovered a superior cyclization method.[178] That the photosensitizer DCA in its ground state ($E_{1/2}(-) = -0.89$ V) can serve as an oxidant for the α-amino radical $E_{1/2}(+) =$ ca. -1 V) makes sense from a thermodynamic viewpoint (i.e., $\Delta G^{O}_{SET} < 0$). Other cyanoarenes which can serve as SET photosensitizers for these processes, such as 1,4-dicyanonaphthalene (DCN) and 1,4-dicyanobenzene (DCB), have lower ground-state reduction potentials than DCA ($E_{1/2}(-) = -1.28$ V for DCN and -1.6 V for DCB) and thus should not oxidize α-amino radicals efficiently. Verification of this proposal has come from the observations that the DCN-sensitized and triphenylene-DCB redox cosensitized reactions of silylamino ester **122** (R = CH_2Ph) give mainly the piperidine **124** along with only trace (1–2%) amounts of the pyrrolidine **125** even when

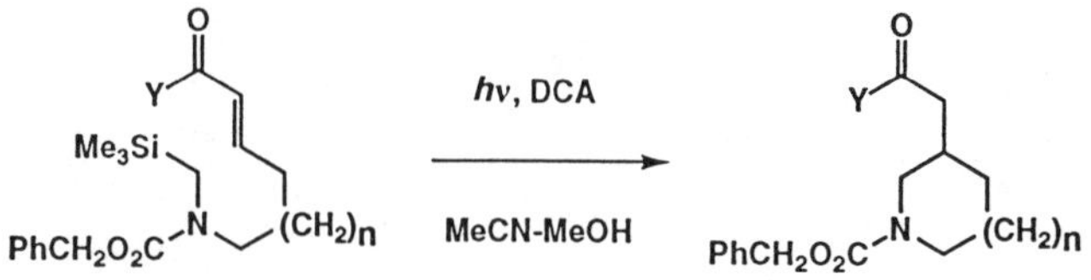

Me_3Si — $PhCH_2O_2C$–N–$(CH_2)_n$ — $h\nu$, DCA / MeCN-MeOH → $PhCH_2O_2C$–N–$(CH_2)_n$

128 (Y=Me or OMe) (n=0 or 1)

the cyanoarenes are present in relatively high concentrations (ca. 10^{-2} M). However, the overall yields of the radical cyclization reactions are low (29–39%) owing to side reactions attributable to the reactivity of these cyanoarenes with α-amino radicals.

A better solution to the problem is found in the use of nitrogen electron–withdrawing substituents to attenuate the oxidation rates of key radical intermediates. Based upon the effects of N-acyl substituents on amine oxidation potentials, Mariano, Yoon, and their co-workers reasoned that the rates of oxidation by DCA of α-silyl-amido and -carbamido radicals would be significantly diminished while both their formation in DCA SET-sensitized photoreactions and their regiocontrolled additions to unsaturated ketones and esters should still be viable. The silylcarbamido-ketones and -esters **128** are indeed converted to the corresponding cyclization products under DCA-photosensitized conditions in high yields (ca. 90%) and to the total exclusion of products arising through α-amido radical oxidation.

5.2.5. Synthetic Aspects of α-Silylamine–Enone SET-Sensitized Photocyclizations

In recent investigations exploring the synthetic potential of silylamine–enone SET-sensitized photocyclization reactions, a number of novel processes have been developed. Reactions of this type included in Scheme 38 demonstrate the applications of this method in which ring formation occurs by both exo- and endoradical cyclization routes, to the synthesis of a variety N-heterocyclic substances.[183,184] In these processes, the α-silylamine or amide functions act as chemical equivalents of the corresponding radical intermediates. The silicon-containing substrates are readily prepared by N-alkylation of the precursor amines with available trimethylsilylmethyl halides. In this way, regiocontrol for

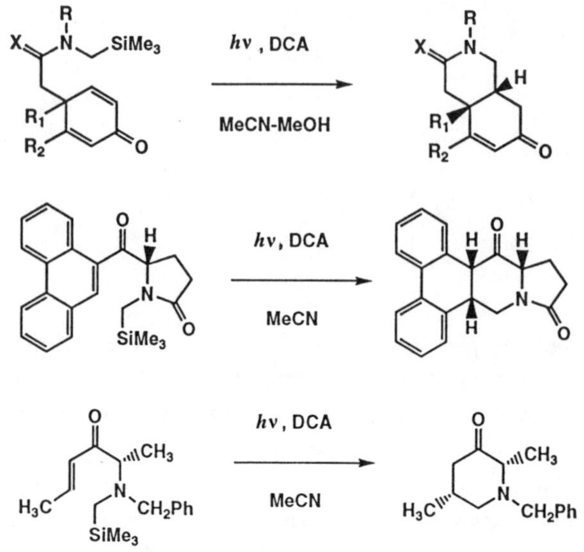

Scheme 38.

α-amino radical generation is established by synthetic manipulation, and it is preserved even when a variety of other, target-governed R-groups are present on nitrogen. Moreover, incorporation of the trimethylsilyl group in the cyclization substrates has another important consequence. Since the products of these photochemical processes are themselves tertiary amines or amides, the oxidative conditions used to promote cyclization could also induce product destruction. The required selectivity preventing this arises from two factors. First, the TMS group causes the oxidation potentials of the reactants to be lower (ca. 0.5 V) than those of the products, and second, the reactivity of the substrate cation radicals ($k_{-TMS^+} >> k_{-H^+}$) is much higher than that of the product cation radicals.

5.3. Mechanistic Studies of α,β-Unsaturated Ketone Photoreduction Reactions Promoted by SET

The SET photoreduction of conjugated ketones is another process mediated by tertiary amines. Intense interest in the relationship of this process to that of excited-state quenching by amines has spanned nearly three decades.[100] Fundamental contributions by Cohen et al.[100] on the mechanism are summarized in Scheme 39. Of the several ketones that

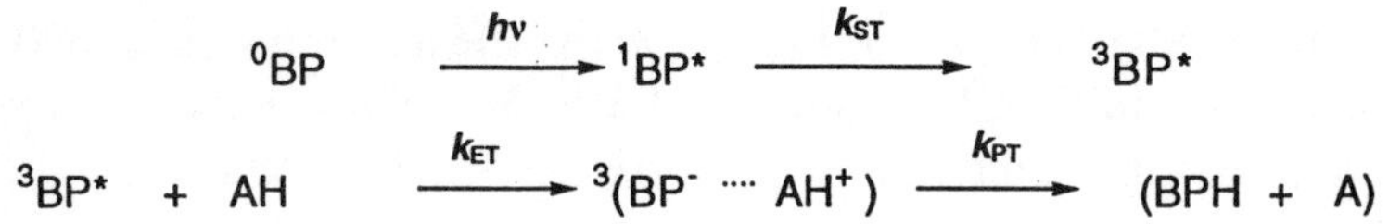

$$^{0}BP \xrightarrow{h\nu} {}^{1}BP^{*} \xrightarrow{k_{ST}} {}^{3}BP^{*}$$

$$^{3}BP^{*} + AH \xrightarrow{k_{ET}} {}^{3}(BP^{-} \cdots AH^{+}) \xrightarrow{k_{PT}} (BPH + A)$$

Scheme 39. The Cohen mechanism[185] for the photoreduction of benzophenone (BP) by amines (AH), where k_{ST} is the rate constant for intersystem crossing (singlet–triplet), k_{ET} is the rate constant for electron transfer, and k_{H+} is the rate constant for proton transfer.

have received attention, benzophenone (BP) has received the most and has become the benchmark for SET reduction of both ketones and enones by a wide variety of electron donors, including amines. In a pioneering study of the time evolution of the BP photoreduction by amines, Simon and Peters[130] employed picosecond transient absorption spectroscopy to observe the primary intermediates. Their results suggested that the initial step in the SET process involves a one-electron transfer from dimethylaniline (DMA) to the triplet state of benzophenone ($^{3}BP^{*}$), initially generating a solvent-separated ion–radical pair (SSIRP) of BP radical anion and DMA radical cation.

Following electron transfer, the SSIRP collapsed to the contact ion–radical pair (CIRP), Scheme 40. Evidence for this process was based primarily on transient absorption spectra following laser excitation (25 ps at 355 nm) of benzophenone and 1.0 M dimethylaniline. Two bands were initially present following excitation. An absorption maximum at 715 nm, assigned to the SSIRP, increased in intensity concomitant with the decrease of the 525 nm band, assigned to $^{3}BP^{*}$. Peters's interpretation was based on the observation of a shift in the absorption maximum from

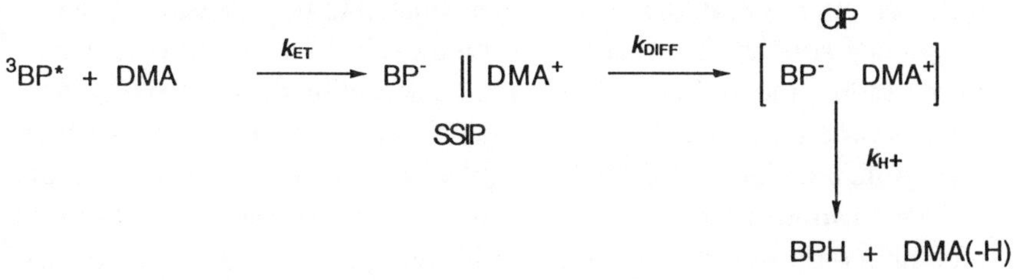

$$^{3}BP^{*} + DMA \xrightarrow{k_{ET}} \underset{SSIP}{BP^{-} \,\|\, DMA^{+}} \xrightarrow{k_{DIFF}} \overset{CIP}{[BP^{-} \; DMA^{+}]} \xrightarrow{k_{H+}} BPH + DMA(-H)$$

Scheme 40. The Peters mechanism[186] for the photoreduction of benzophenone by dimethylaniline.

715 nm to 690 nm over a period of 200 ps. The new transient at 690 nm was assigned to the CIRP. Small bathochromic shifts in the absorption maxima on solvent interdigitation into CIRP's to form SSIRP's are well documented.[1]

Similar time-dependent shifts in the absorption spectra were observed for the photoreduction of benzophenone with 5-M DMA in MeCN, with 1-to-5-M diethylaniline (DEA), and with 1-M dabco in solvents that ranged in polarity from benzene to MeCN. A subsequent rapid protonation that formed the ketyl radical, λ_{max} at 545 nm, was shown to be kinetically coupled to the 690-nm transient but not the 720-nm transient. The rate constant for proton transfer was determined to be 2.0×10^9 s^{-1} in 1.0-M DMA. The relative ratios of radical ion to ketyl radical absorbance were assumed to be a measure of the ratio of CIRP to ketyl radical. The ratio decreased with solvent polarity from MeCN to hexanenitrile in accord with the expected polarity of the transients.

Subsequently, Manring and Peters[129] further probed the mechanism for proton transfer within the ion–radical pair (IRP) formed in the photoreduction of benzophenone by diphenylmethylamine (DPMA) by monitoring the time-dependent transient absorption spectrum of the DPMA radical cation at 635 nm. The decrease of the 635-nm absorption was ascribed to proton transfer within the CIRP. Both solvation of the CIRP and charge recombination were ruled out as contributors to the decay of the transient. The rate constant for proton transfer from the DPMA radical cation was greater than 5×10^9 s^{-1} for a series of solvents ranging in polarity from benzene to methylene chloride, whereas that for dissolution to the SSIRP was estimated to be less than 2×10^6 s^{-1} based on (1) the known association constants for ion pairs in solvents with dielectric constants similar to those used in this study ($\varepsilon \approx 12$), and (2) the assumption that the rate for association of an ion pair is expected to be diffusion limited ($k_{diff} \approx 1 \times 10^{10}$ M^{-1} s^{-1}). The rate constant for charge recombination within the CIRP was estimated to be at least an order of magnitude smaller than that for proton transfer based on kinetic deuterium isotope studies. Peters[130] observed that proton transfer from DPMA radical cation to 4-chlorobenzophenone radical anion was faster than deuterium transfer from DPMA-d_3. Yet at 50 ns the yield of ketyl radicals from both amines was the same. Competitive proton transfer and charge recombination would have required different yields of the ketyl radicals.

Devadoss and Fessendon[186] challenged Peters's assertion that the SSIRP was the intermediate formed upon electron transfer. In their studies[186,187] the photoreduction of benzophenone by several tertiary amines was examined using picosecond and nanosecond spectroscopy. Excitation of 0.01-M BP with 0.1-to-0.5-M dabco in benzene, MeCN, or MeOH led to formation of a transient with an absorption maximum at 525 nm, assigned to $^3BP^*$. Decay of the $^3BP^*$ band occurred concomitant with formation of transients with absorption maxima at 725 nm in benzene, 720 nm in MeCN, and 615 nm in MeOH. The blue shift of the absorption maximum was attributed to H-bonding of the BP radical anion. For 1-M dabco, a small shift from 700 nm to 720 nm with increased solvent polarity was observed, opposite in direction to that reported by Peters.[130] In a reversal from Peters's assignment of the transients, Fessendon attributed the bands at 700 nm and 720 nm to the SSIRP and CIRP, respectively, based on the assumption that the SSIRP would be lower in energy than the CIRP.[185]

The lifetimes of the ion pairs were also solvent dependent. In nonpolar solvents or in π-donor solvents, the IRP was shorter lived and the ketyl radical yield was higher; in benzene, for example, the IRP had a lifetime of 50 ns and a ketyl yield of 80% whereas in CH_3CN, the lifetime was 2.1 μs with a ketyl yield of 20%. In contrast to the DMA and DPMA studies, the mechanism for photoreduction of BP by dabco was proposed to be electron transfer from dabco to $^3BP^*$ to form the CIRP directly. The decay pathways from the CIRP in nonpolar solvents were charge recombination and proton transfer. In more polar solvents, separation of the CIRP into a SSIRP dominated.

The decay of the BP–dabco CIRP was reported to be first order, even in highly polar solvents where ion pair dissociation would be expected. Moreover, only a slight increase in the ion pair lifetime was observed on "substantially" reducing the laser power. First-order decay of the CIRP in MeCN had also been reported by Inbar, Linschitz, and co-workers.[101]

These interpretations were, in turn, challenged by Miyasaka and co-workers.[188] Under experimental conditions similar to those of Devadoss and Fessendon, photolysis of 0.01-M BP and 0.3-M dabco in MeCN with a 22-ps (355-nm) pulse produced initially $^3BP^*$ (525 nm), which decayed concomitant with formation of the BP radical anion, reaching a concentration plateau within 1 ns. Parallel photoconductivity studies indicated that a maximum free-ion current was also attained within 2 ns. Based on these two observations, the rate for ion pair dissociation was estimated to be $> 0.5 \times 10^9\ s^{-1}$. The decay of the band at 700 nm occurred by mixed-order kinetics, second-order for the first 1.5 μs then, at longer times, a mixture of first- and second-order processes.

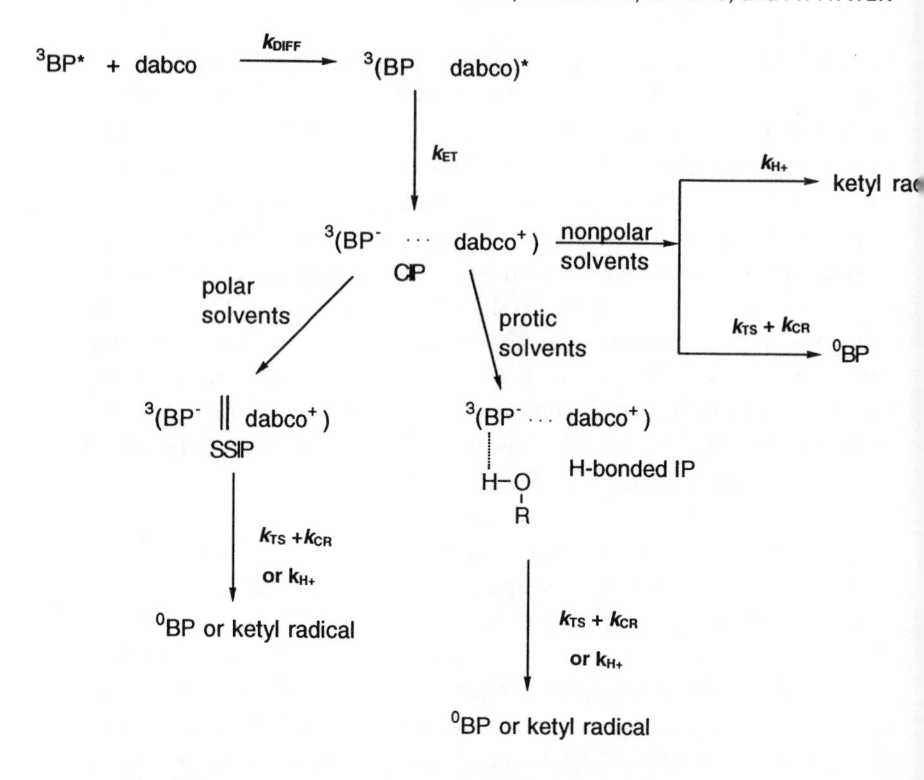

Scheme 41. Fessendon's mechanism[189] for the photoreduction of benzophenone by dabco, where k_{DIFF} is the bimolecular rate constant for diffusion, k_{ET} is the rate constant for electron transfer, k_{H+} is the rate constant for proton transfer, k_{TS} is the rate constant for intersystem crossing (triplet–singlet), and k_{CR} is the rate constant for charge recombination.

At concentrations between 0.1 M and 1.0 M dabco in MeCN, both the decay of BP radical anion and the yields of ketyl radical increased with increasing dabco concentration. The time-dependent decay of the absorption of the BP radical anion was in accord with the reaction mechanism given in Scheme 42, which included second-order decay for recombination of BP radical anion and dabco radical cation, first-order decay for BP radical anion and ground-state dabco, and zero-order decay of BP radical by direct decomposition or impurity scavenging of the radical. The experimental data correlated well with this mechanistic model, with a second-order rate constant k_2 of 1.74×10^{10} M^{-1} s^{-1} for the diffusion-controlled radical anion–radical cation annihilation step (k_{diff} = 1.8×10^{10} M^{-1} s^{-1}). The plot of k_1 [dabco] versus [dabco] was linear

$$^{3}BP^{*} + dabco \xrightarrow{k_{ET}} {}^{3}(BP^{-\cdot} \cdots dabco^{+\cdot})$$

$$^{3}(BP^{-\cdot} \cdots dabco^{+\cdot}) \xrightarrow{k_{ID}} BP^{-\cdot} + dabco^{+\cdot}$$

$$BP^{-\cdot} + dabco^{+\cdot} \xrightarrow{k_2} BP + dabco$$

$$BP^{-\cdot} + dabco^{+\cdot} \xrightarrow{k_1} BP^{\cdot} + dabco$$

$$BP^{-\cdot} \xrightarrow{k_0} X^{-}$$

Scheme 42. Mataga's mechanism[192] for the photoreduction of benzophenone by dabco, where k_{ET} is the rate constant for electron transfer, k_{ID} is the rate constant for ionic dissociation, k_2 is the bimolecular rate constant for charge annihilation, k_1 is the rate constant for proton transfer from dabco, and k_0 is the rate constant for proton transfer from solvent impurities.

and yielded $k_0 = 3.0 \times 10^5$ s^{-1} and $k_1 = 4.8 \times 10^5$ M^{-1} s^{-1}. Similar conclusions were reached by Haselbach et al.[189] and Peters and Lee.[190]

In a subsequent study, Devadoss and Fessendon[187] examined the photoreduction of benzophenone by diethylaniline (DEA) and triethylamine (TEA). Excitation of 0.010 M BP with 0.02 M DEA in benzene with an 18-ps pulse at 355 nm led to formation of a transient ion pair with an absorption maximum at 735 nm, assigned to the BP radical anion, and a transient at 470 nm, assigned to the DEA radical cation. The half-life of the IRP was 875 ps. The IRP decayed to the ketyl radical with a yield of 85%.

In MeOH the formation of transients assigned to the BP–DEA radical ion pair was observed under similar conditions. In contrast to the photoreduction in benzene, in which the IRP decayed with first order kinetics, in methanol the loss of the radical anion was first order while the loss of the radical cation was second order. When MeOD was the solvent, a k_H/k_D of 1.84 was obtained, indicating that proton transfer occurred from the hydroxyl group of MeOH.

Photoreduction in MeCN under the same experimental conditions also led to formation of the ion radical pair and ketyl radicals. The second-order decay of the IRP did not result in an increase in the ketyl radical absorption, the yield of which remained unchanged at 20%. When the concentration of DEA was increased to 1.0 M, the BP radical's absorption peak shifted from 702 nm to 727 nm during the first 50 ps and the BP radical anion and DEA radical cation formed at an earlier time. The time-dependent shifts were attributed to the formation of a DEA–DEA radical cation dimer. The proposed mechanism for this photoreduction is similar to that proposed by the authors for the BP photoreduction by dabco (Scheme 43). Proton transfer within the CIRP dominates in nonpolar solvents. In polar solvents, the CIRP decays to a SSIRP or separate ions prior to proton transfer. A theoretical treatment[190] showed that CIRP formation between benzophenone radical anion and diethylaniline radical cation is favored in benzene and cyclohexane, while

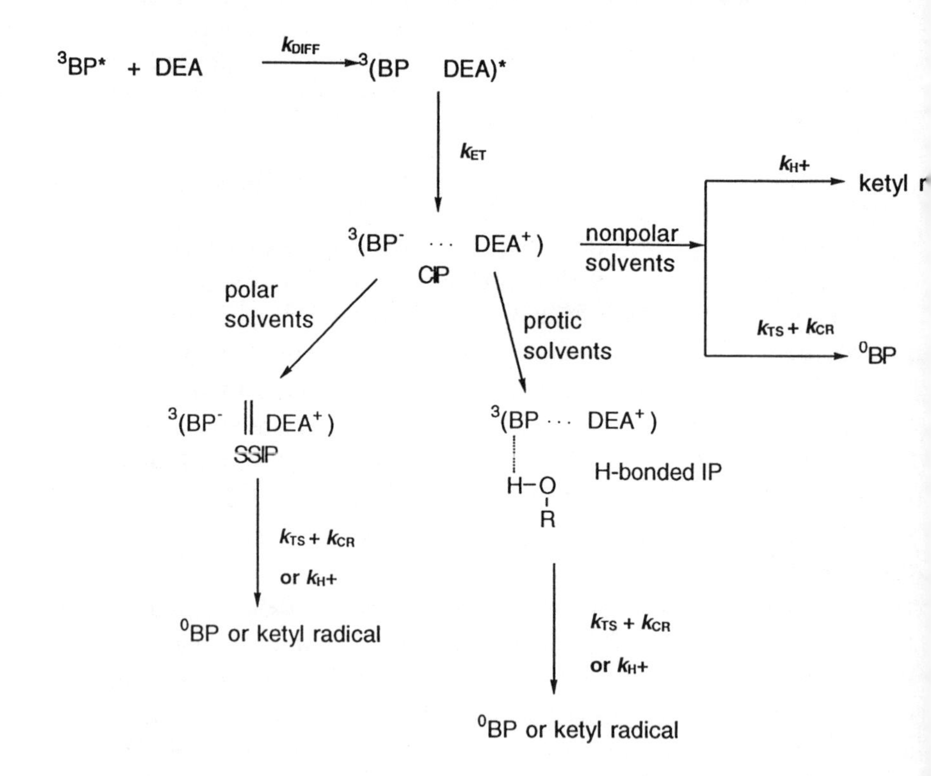

Scheme 43. Fessendon's mechanism[180] for the photoreduction of benzophenone by dabco, where k_{DIFF} is the bimolecular rate constant for diffusion, k_{ET} is the rate constant for electron transfer, k_{H+} is the rate constant for proton transfer, k_{TS} is the rate constant for intersystem crossing (triplet–singlet), and k_{CR} is the rate constant for charge recombination.

SSIRP was favored in *o*-dichlorobenzene, acetone, MeOH, and MeCN. Unfortunately, the energies for the IRP used in the calculations were not given.

Aliphatic amines were also incorporated in this study. When TEA was employed as the donor in MeOH, MeCN, and benzene, no direct evidence for the IRP was observed. A band at 600 nm that decayed in less than 20 ps was assigned tentatively to the IRP. However, indirect evidence (see below) for electron transfer followed by proton transfer was obtained. The value for ΔG_{CIRP} was calculated to be –0.30 eV in MeCN and MeOH and –0.11 eV in benzene. The larger quenching-rate constants were assumed to indicate that hydrogen abstraction was not the first step, $k_Q = 8 - 32 \times 10^8$ M^{-1} s^{-1} for these solvents in comparison with an experimental rate constant for hydrogen abstraction by $^3BP^*$ in 2-propanol of 1.8×10^6 M^{-1} s^{-1}.[7] The quenching rate constants of the triplet state for 4,4′-dichlorobenzophenone, 4-chlorobenzophenone, and 4,4′-dichlorobenzophenone by TEA were estimated to be higher than those for TEA quenching of $^3BP^*$. In fact, hydrogen abstraction is expected to be even slower with these substrates.

The best evidence[187] for a SET pathway with TEA resulted from the photoreduction of BP in MeOH. The slow quenching rate in MeOH implied that a hydrogen-bonding interaction was reducing the electron transfer rates. In the more acidic trifluoroethanol, the rates were even lower (e.g., $k_Q = 1.8 \times 10^7$ M^{-1} s^{-1}). This was attributed to hydrogen bonding to the amine, increasing the n (amine) – π^* (BP) energy gap.

By this point much direct evidence supporting SET as the initial step of amine photoreduction of enones had been acquired. Efficient proton transfer within the ion pair followed the electron transfer. The nature of the ion pair, whether contact or solvent-separated, was still a mystery. Insight into this puzzle was provided by a series of studies undertaken by Mataga and co-workers. In 1990, Miyasaka and Mataga[192,193] reported the photoreduction of benzophenone with 0.05 M DMA in MeCN at 355 nm with a 22-ps pulse. The initially generated $^3BP^*$ decayed at a rate equal to that for the formation of the benzophenone ketyl radical, DMA radical cation and BP radical anion. No time-dependent spectral shift of the 700 nm band was observed. Furthermore, time-resolved photoconductivity measurements indicated that the formation of free solvated ions occurred within a few nanoseconds, the time resolution of the instrument. The time profiles of each of the transients were fitted successfully to the mechanism shown in Scheme 44. The efficiencies of ionic dissociation

$$^3BP^* + DMA \xrightarrow{k_{ET}} {}^3(BP^- + DMA^+)_{enc}$$

$$^3(BP^- + DMA^+)_{enc} \xrightarrow{k_{ID}} BP^- + DMA^+$$

$$^3(BP^- + DMA^+)_{enc} \xrightarrow{k_{H+}} BP^\cdot + DMA^\cdot$$

Scheme 44. Mataga's mechanism[188] for the photoreduction of the benzophenone by dimethylamine, where k_{ET} is the rate constant for electron transfer, k_{H+} is the rate constant for proton transfer, and k_{ID} is the rate constant for ionic dissociation.

and ketyl radical formation, calculated from the time/concentration profiles of the transients, were 0.17 and 0.73, respectively.

At amine concentrations as high as 1.0 M DMA, the absorption spectral shifts evolved in a manner similar to those observed with 0.05 M DMA except for the rapid increase in the absorbance attributed to the IRP (Table 5). The experimental time-transient profiles were also successfully fitted to the mechanism outlined in Scheme 44.

The initial increase in absorbance from the IRP seen at high amine concentrations (> 0.10 M) probably involved contributions from either static quenching of singlet-state benzophenone or direct excitation of a ground-state BP–DMA complex in competition with the bimolecular quenching of $^3BP^*$ by the amine. Support for the direct excitation pathway came from the observation of an increase in absorbance and slight red-shift in the onset of the BP absorption possibly because of a weak ground-state complex at high DMA concentrations. The equilibrium constant for the complex was estimated to be between 0.1 and 0.5 M^{-1} by Benisi–Hildebrand treatment. Excitation of the ground-state complex (397 nm) with a picosecond pulse led after 1 ps to the BP radical anion. The first-order decay of the absorption gave a lifetime of 85 ps.

Table 5. Lifetimes of the IP Complexes and Rate Constants for Proton Transfer (k_{H^+}), Ionic Dissociation (k_{ID}), and Charge Recombination (k_{CR})

IRP complex	τ *(ps)*	k_{H^+} (s^{-1})	k_{ID} (s^{-1})	k_{CR} (s^{-1})
$^3(BP^- \cdots DMA^+)$enc	140	5.4×10^9	1.4×10^9	—
$^1(BP^- \cdots DMA^+)$enc	460	6.6×10^8	9.5×10^8	5.6×10^8
$^1(BP^- \cdots DMA^+)$com	85	$<<10^8$	$\leq 4 \times 10^8$	1.1×10^{10}

The BP ketyl radical was not observed under these conditions; thus charge recombination dominated the decay route of the excited complex $^1(BP^{-\cdot} \cdots {\cdot}DMA^{+})_{com}$.

The role of static quenching of $^1BP^*$ by DMA nearest neighbors was explored by following the time-dependent decay of the $S_n \leftarrow S_1$ absorption spectrum of benzophenone as a function of DMA concentration. Excitation of 0.05-M benzophenone in MeCN at 355 nm with a 500-fs pulse led at 1 ps to a spectrum with an absorption maximum at 575 nm assigned to $^1BP^*$. The decay of this transient led to formation of $^3BP^*$ with a rise time of 9 ps. At 1.0 M DMA, the absorption spectrum at 1 ps following excitation showed bands with maxima at 470 nm (DMA^+) and 575 nm ($^1BP^*$) and a broad absorption at $\lambda_{max} > 600$ nm (BP^-), which increased in intensity. The very rapid formation of $^1(BP^{-\cdot} \cdots {\cdot}DMA^{+})_{com}$ was attributed to excitation of the weak ground-state complex. At 1.0 M DMA in MeCN, the efficiencies for $^1(BP^{-\cdot} \cdots {\cdot}DMA^{+})_{com}$ and $^1BP^*$ were 0.30 and 0.70, respectively. At this concentration of DMA, the lifetime of $^1BP^*$ was 5.0 ps. Half of the $^1BP^*$ was quenched by DMA, and half underwent intersystem crossing to $^3BP^*$. The bimolecular quenching rate constant of $1.1 \times 10^{11}\ M^{-1}\ s^{-1}$ is approximately five times larger than the rate constant for diffusion, which was attributed to nearest-neighbor quenching of $^1BP^*$.

With the efficiencies for the initial formation of $^1(BP^{-\cdot} \cdots {\cdot}DMA^{+})_{com}$ and the time-dependent rate profiles for photoreduction of BP in 1.0 M DMA, the yield and rate constants for proton transfer (k_{H^+}), ion pair dissociation (k_{ID}), and charge recombination (k_{CR}) from $^1(BP^{-} \cdots {\cdot}DMA^{+})_{enc}$ were determined and are listed in Table 5. Scheme 45 summarizes the Mataga mechanism for the photoreduction of BP by DMA.

Miyaska and co-workers[193] extended the examination of the photoreduction of benzophenone with *N,N*-diethylaniline (DEA), *N,N*-diethyl-*p*-toluidine (DET), and *N*-methyldiphenylamine (MDA) in order to evaluate the relationship between ΔG_{ET} and the properties of the CIRP complexes. Indeed, in contrast to Devadoss and Fessendon,[187] these authors[193] were able to incorporate direct ketyl formation from the CIRP into their mechanistic interpretation.

A weak ground-state charge-transfer complex was also formed for BP with high DEA concentrations that upon excitation produced a short-lived singlet-state ion pair. However, unlike $^1(BP^{-\cdot} \cdots {\cdot}DMA^{+})_{com}$, for dimethylaniline, one tenth of the DEA singlet complex, $^1(BP^{-} \cdots {\cdot}DEA^{+})_{com}$ dissociated into free ions. The remaining 90% decayed by

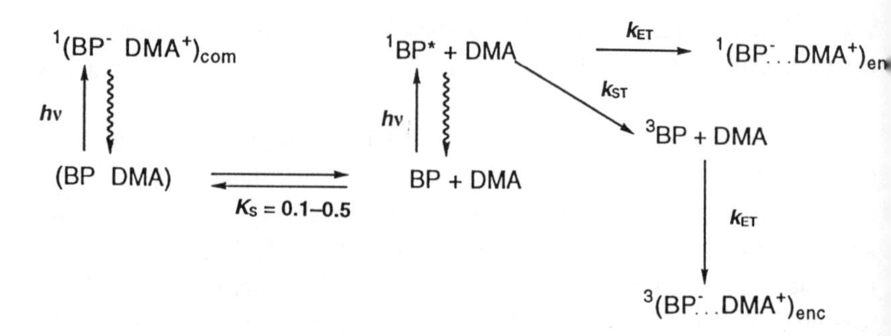

Scheme 45. Mataga's summary[188] of the excited-state reactions in the photoreduction of benzophenone by dimethylamine, where k_{ET} is the rate constant for electron transfer, k_{ST} is the rate constant for intersystem crossing (singlet–triplet), and K_s is the association constant for benzophenone and dimethylaniline.

way of charge recombination. The E_{OX} affected the decay lifetimes and proton transfer rates from the triplet state IRP (summarized in Table 6), whereas the proton transfer rate constant from singlet IRP_{enc} showed little variation with E_{OX}. All four ion pairs had K_{H^+} of 5–8 × 10^8 s^{-1}.

Much of the discussion in the work reviewed here has been centered on the structure of the IRP formed by electron transfer from amine to benzophenone. Mataga[192] and Peters[130] both proposed that electron transfer occurs upon diffusive encounter of a tertiary aromatic amine and $^3BP^*$ to form a SSIRP. Mataga further suggested that excitation of ground-state amine–BP CT complex led directly to a "tight" ion–radical pair that decayed mainly through charge annihilation. The interionic distance of the ion–radical pair formed by interception of $^1BP^*$ by an amine, followed by electron transfer, is intermediate between those for $^3(BP^- \cdots DMA^+)_{enc}$ and $^1(BP^- \cdots DMA^+)_{com}$. These hypotheses depend on a

Table 6. Properties of the Radical–Ion Pairs Resulting from Bimolecular Quenching of $^3BP^*$, Determined from Fitting the Time Profiles of the Transients to Scheme 44

	MDA	*DMA*	*DEA*	*DET*
E_{OX} (V versus SCE)	0.86	0.76	0.72	0.69
τ, (^{3}IP)enc (ps)	90	140	290	>500
k_{H^+} (s^{-1})	8.0 × 10^9	5.4 × 10^9	7.3 × 10^8	<<2 × 10^8
k_{ID} (s^{-1})	1.1 × 10^9	1.4 × 10^9	2.1 × 10^9	2 × 10^9

geometry of the transition state for proton transfer that is more restrictive than that for electron transfer.

Manring and Peters[129] showed that the ion–radical pair formed by photoinduced electron transfer from *N*-methylacridan to benzophenone decayed with a half-life of 500 ps to yield the BP ketyl radical, λ_{max} = 545 nm, and the NMA radical cation, λ_{max} = 520 nm. The isotope effect on the proton transfer was explored with 9,9-dideuterio-*N*-methylacridan, which yielded $k_H/k_D = 1.00 \pm 0.03$, as measured by decay of the IP, indicating no change in the rate. However, for *N*-methylacridan-methyl-d_3 as the electron/proton source, an isotope effect of $k_H/k_D = 1.4 \pm 0.05$ was obtained, indicating that the proton transfer occurred from the *N*-methyl group rather than from the 9-methylene position. The proton abstraction of the kinetically favored proton from the *N*-methyl group over the thermodynamically favored 9-proton was cited as evidence for an IRP geometry of the BP–NMA ion–radical pair in a coplanar head-to-head orientation. This conformation has maximum π overlap and the shortest interionic distance. Proton migration from the *N*-methyl group occurs by a least-motion transfer within the ion pair.

This same study measured deuterium kinetic isotope effects on the rate of proton transfer in the ion pairs formed between DPMA/DPMA-d_3 for BP, 4-chlorobenzophenone, and 4,4′-dichlorobenzophenone. The primary kinetic isotope effect (PKIE) k_H/k_D, measured from the rate of disappearance of $^3(BP^{+\cdot} \cdot\cdot \cdot DPMA)_{enc}$, varied from 1.9 in benzene to ≈ 2.5 in chlorobenzene, pyridine, and methylene chloride and was much smaller than expected for a symmetrical transition state found when $\Delta G^\circ = 0$.[10,11] The PKIE in chlorobenzene ($E_T = 37.5$) and in methylene chloride ($E_T = 41.1$) were essentially equivalent even though the process is less exergonic in the latter, more polar solvent. A lower-than-expected PKIE could be ascribed to solvent reorganization prior to proton transfer. If the contribution to the transition-state free energy due to solvent reorganization approaches that due to the proton-transfer step itself, then the PKIE could not be as significant a measure of the process since a minimal isotope effect is expected for IRP reorganization. Peters[129] tested this hypothesis with chlorine-substituted benzophenones that form more stable radical anions and thus place a lower demand on the solvent reorganization. The values of k_H/k_D for BP, 4-chlorobenzophenone, and 4,4′-dichlorbenzophenone were 1.9, 2.8, and 3.0, respectively. For these three benzophenones, the effect of stabilization was expected to be greater on the transition-state energy for proton transfer than on IRP-solvent reorganization, consistent with the observed trend in the isotope effect. Wagner[191] had suggested in his earlier studies on the Type-II photoreactions of α, β, and γ-dialkylaminoketones that geometrical constraints limited proton transfer rates.

Mataga and co-workers[192] also evaluated the geometrical constraints on ion pairs that control the rate of proton transfer. As noted above, the Mataga ion pairs were formed with differing interionic distances depending on their mode of production. As shown in Table 5, the three decay pathways reveal two general patterns: (1) rate constants for proton transfer and ion-pair dissociation decrease across the series $^{3}(IRP)_{enc}$, $^{1}(IRP)_{enc}$, $^{1}(IRP)_{com}$, whereas (2) the rate of charge recombination k_{CR} increases. Ion-pair dissociation was presumed to be most facile from the SSRIP and, consequently, the rate of ion-pair dissociation to free ions (k_{ID}) was measured to be largest from $^{3}(BP^{-\cdot}\cdots DPMA^{-})_{enc}$. Charge recombination should be fastest from the CIRP and, indeed, was faster for the IRP formed by excitation of the ground-state complex, $^{1}(BP^{+\cdot}\cdots DPMA^{-})_{com}$. The proton-transfer rate constant was largest from $^{3}(BP^{+\cdot}\cdots DPMA^{-})_{enc}$. If the 720-nm transient were the SSIRP, then proton transfer would have to occur over a relatively long distance (Figure 1). Using van der Waal's radii, the center-to-center distance for the IRP with one intervening acetonitrile molecule is 10.8 Å. One would normally expect proton transfer to be fastest over a shorter distance. The lower rate for proton transfer within $^{1}(BP^{-\cdot}\cdots DPMA^{+})_{enc}$ was attributed to a slower reorganization rate for the CIRP. As noted above, Peters[190] also invoked slow IRP reorganization to explain proton abstraction from the kinetically favored position instead of the thermodynamically favored position in *N*-methylacridan. In contrast to Peters's proposed mechanism, however, interconversion between CIRP and SSIRP is not required for the Mataga mechanism, even for relatively long-lived ion pairs such as $^{1}(BP^{-\cdot}\cdots DPMA^{+})_{enc}$ ($\tau = 460$ ps). Using simple diffusion theory and assuming a diffusion coefficient for BP in MeCN of 1×10^{-5} cm^{2} s^{-1}, BP will diffuse 3 Å, the diameter of one solvent molecule, in 90 ps. As noted above, the lifetime of $^{1}BP^{*}$ is only a few picoseconds. If interconversion between SSIRP and CIRP were this fast, then one would not expect a difference in the lifetimes of $^{1}(BP^{-\cdot}\cdots DPMA^{+})_{enc}$ and $^{3}(BP^{-\cdot}\cdots DPMA^{-})_{enc}$ to be 320 ps.

Mataga[193] has also studied the SET photoreduction of benzophenone as a function of the amine oxidation potential. In Table 6, the lifetimes of $^{3}(IRP)_{enc}$ with different tertiary amines and the rate constants for proton transfer and ion-pair dissociation are listed as functions of the amine oxidation potential E_{OX}. The lifetimes and rate constants of

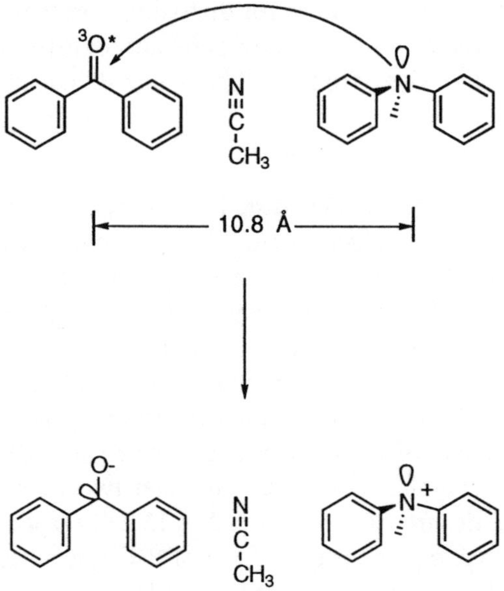

Figure 1. Long-distance electron transfer from $^3BP^*$ to DPMA.

ion-pair dissociation vary in direct proportion to the amine oxidation potential E_{OX}, whereas the proton-transfer rate constant is inversely proportional to E_{OX}. The free energy for electron transfer ΔG_{ET} also decreases with E_{OX}. Mataga suggests that as ΔG_{ET} becomes more negative, the distance at which electron transfer takes place should increase. Thus the interionic distance in the RIP formed between BP and DET (E_{OX} = 0.69 V) is larger than that for BP and MDA (E_{OX} = 0.86 V). The small value for k_{H+} in $^3(BP^+ \cdots DET^-)_{enc}$ is presumed to result from the large interionic distance in the IRP and the inability of the IRP to collapse to a geometry suitable for proton transfer within the lifetime of the complex, $\tau > 500$ ps. One would expect that if electron transfer were taking place over a larger distance for donor–acceptor pairs with more negative ΔG_{ET}, then the observed rate constant for loss of $^3BP^*$ would also increase. The observed rate constant for disappearance of $^3BP^*$ is given by

$$k_{OBS} = \frac{k_{ET} \times k_{DIFF}}{k_{ET} + k_{-DIFF}}$$

where k_{DIFF} is the rate constant for diffusion together of the donor–acceptor pair and k_{-DIFF} is the rate constant for diffusion apart of the encounter pair and where from the Smoluchowski equation,[194,195]

$$k_{DIFF} = 4\pi\rho D$$

where ρ is the intermolecular distance between the members of the donor–acceptor pair at reduction and D is the mutual diffusion coefficient for the donor and the acceptor. The value of ρ for electron transfer would increase as the electron transfer distance increases, and consequently so would k_{OBS}. From the time profiles for the loss of $^3BP^*$ with 0.3-M DMA[196] and 0.3 M DEA,[197] in acetonitrile, the calculated rate is the same within 10%. The lifetimes for the ion pairs vary by a factor of two, while k_{ID} varies by a factor of seven. Once again using van der Waal's radii for the BP–DMA complex with zero, one, and two intervening solvent molecules, ρ varies from 6.4 Å to 10.8 Å to 15.2 Å, respectively. A better test would be to compare the rate of loss of $^3BP^*$ quenched by MDA and DET. Unfortunately, those data are not available.

The influence of geometry on proton transfer was at the heart of a study by Miyasaka et al.,[198] who further examined the intramolecular photoreduction of BP. In the study, 4-oxybenzophenone was linked to 4-oxydiphenylamine by either a two- or a three-methylene bridge spacer, giving structures BO2OD and BO3OD, respectively. The distance between the carbonyl oxygen of BP and the N-H hydrogen of the diphenylamine was calculated using the weighted Boltzmann factor for each possible conformation.The distribution of conformers for BO2OD was skewed with NH–O distances of between 9 Å and 15 Å. The peak of the population distribution was at 10.5 Å. The conformer distribution for BO3OD had two bands. One was skewed from 4–7.5 Å with a population maximum of 25% at 6.5 Å, and the other was skewed from 10–16.5 Å with a maximum of 75% at 13.5 Å. The shortest O–H distance for BO3OD was 5 Å, while the shortest O–H distance for BO2OD was 9 Å.

Photolysis of BO3OD at 355 nm with a 22-ps pulse resulted in a 530-nm transient ($^3BP^*$), which decayed concomitant with formation of the BP ketyl radical (555 nm) and the DMA cation radical (690 nm). This latter absorption also contained contributions from the 700-nm BP anion. In contrast to the tertiary aromatic amines, the time profiles of the transients from these amines could not be fitted to a mechanism in which direct hydrogen abstraction k_{HA} of the amine hydrogen from $^3BP^*$ competed with electron transfer k_{ET} followed by ionic dissociation (Scheme 46). However, the profiles could be fitted to a combination of the two

$$^3BP^* - DPA \xrightarrow{k_{HA}} BP^{\cdot} - DPA^{\cdot}$$

$$^3BP^* - DPA \xrightarrow{k_{ET}} {}^3(BP^{\cdot -} - DPA^{+})$$

$$^3(BP^{\cdot -} - DPA^{+}) \xrightarrow{k_{LL}} BP^{\cdot -} - DPA^{+}$$

Scheme 46. Mataga's mechanism[199] for the photoreduction of the benzophenone–diphenylamine conjugate, where k_{ET} is the rate constant for electron transfer, k_{HA} is the rate constant for hydrogen abstraction from diphenylamine, and k_{LL} is the rate constant for the formation of a long-lived ion pair.

mechanisms. The time-dependent transients for both BO2OD and BO3OD were fitted to the same hybrid mechanism as for BO3OD. The results from the fit of the time profiles are listed in Table 7.

The differences between the two tethered systems and the bimolecular photoreduction of BP by tertiary aromatic amines were attributed to the geometrical constraints on proton transfer and hydrogen abstraction, which have much stricter requirements than the electron transfer process. For example, the triplet state decay of BO2OD by electron transfer is faster than that of BO3OD because of the shorter average intramolecular donor–acceptor distance for the former. For proton transfer or direct hydrogen abstraction, the donor–acceptor distance between the proton and the BP carbonyl must approach the sum of the OH and NH bond

Table 7. Results of Analysis of Time Profiles of Transients from Excitation of BO2OD and BO3OD to a Hybrid Mechanism for Hydrogen Abstraction k_{HA} and Electron Transfer k_{ET}

	BO3OD	*BO2OD*
Lifetime of $^3BP^*$ (ns)	1.75	0.95
Yield of direct HA	0.30	0.07
Yield of stable IRP formation	0.35	0.65
Lifetime of stable IRP (ns)	2.75	9.55
Yield of intra-IRP PT	0.90	1.0
Yield of long-lived IRP	0.18	0.16
Contribution of PT in the stable RP to all of the photoreduction	0.51	0.90

$$^{3}BP^{*} - DPA \xrightarrow{k_{ET}} {}^{3}(BP^{\cdot -} - DPA^{+})$$

$$^{3}(BP^{\cdot -} - DPA^{+}) \xrightarrow{k_{LL}} BP^{\cdot -} - DPA^{+}$$

$$^{3}BP^{*} - DPA \xrightarrow{k_{HA}} BP^{\cdot} - DPA^{\cdot}$$

Scheme 47. Mataga's mechanism[194] for the photoreduction of the benzophenone–diphenylamine conjugate, where k_{ET} is the rate constant for electron transfer, k_{HA} is the rate constant for hydrogen abstraction from diphenylamine, and k_{LL} is the rate constant for the formation of a long-lived ion pair.

lengths. BO3OD can achieve a conformation with a shorter donor–acceptor distance which leads to a rate constant for intra-IRP proton transfer that is larger for BO3OD than it is for BO2OD. In benzene, the low polarity of the solvent encourages formation of a tight IRP. Excitation in benzene resulted in the triplet-state lifetimes for BO3OD of 9.0 ns and for BO2OD of 15.2 ns. The authors speculate that BO3OD is already in a conformation in which there is good overlap between the two π systems of the donor–acceptor pair and that proton transfer is rapid, whereas in BO2OD such a conformation is less easily achieved.

From the foregoing discussion, it is clear that there are many common features to the photoreduction of BP by tertiary amines. Diffusive encounter of a tertiary aromatic amine with $^{3}BP^{*}$ results in electron transfer to give an ion–radical pair. The decay paths available to the relatively short-lived ion pair ($\tau < 2$ ns) include proton abstraction, charge recombination, and ion-pair dissociation. The structure of the ion pair and, indeed, the nature of the ca.-720-nm transient, which originally was assigned to BP radical anion, are still uncertain. Devadoss and Fessendon[186] concluded that electron transfer occurs on every encounter of the amine with $^{3}BP^{*}$ and have assigned the 720-nm band to a CIRP. Peters[130] and Mataga[192,193] suggested that the electron transfer process occurs over a much larger distance and have assigned the 720-nm band to a SSIRP. Quenching rate constants for long-range electron transfer should be higher than substrate-encounter diffusion rate constants because of the increase in p. Indeed, Mataga[192] and Peters[190] in recent studies have reported larger rate constants for electron donor–acceptor pairs that have ΔG_{ET}'s in the Marcus inverted region. An intriguing proposal by Mataga involves transfer over long distances through a structured SSIRP. It is a testament to the richness and complexity of SET reductions of enones that a thorough understanding of the fundamental processes remains elusive.

5.4. The Chemistry of Semi-Enones formed by SET to α,β-Unsaturated Ketones

Models for the chemical processes that emanate from intermediates generated by single-electron transfer from an amine to an excited triplet state of an α,β-unsaturated ketone are readily conceived. A semi-enone radical anion intermediate produced in this manner would be expected to behave analogously, both chemically and physically, to radical anions formed in dissolved metal[201] and electrochemical SET reduction reactions.[202] Furthermore, for α,β-unsaturated ketones, ECE (*e*lectron transfer–*c*hemistry-*e*lectron transfer)[202] reactions are central to the understanding of the chemical consequences in a number of electrochemical and dissolved-metal reduction reactions.[201] In a pivotal study of the Birch reduction of enone **129** (Scheme 48), Stork[203] demonstrated that the β-carbon of the semi-enone radical anion **132** ($X = OTs$) exhibits nucleophilic character through intramolecular displacement of a tosylate group located two carbons from the β-carbon (Scheme 49). On displacement, the propellanone **130** was formed. Subsequently, the β-carbon nucleophilicity has also been demonstrated by Gassman[205] in studies on the electrochemical reduction of **129**. This report was followed by a study of the one-electron transfer from lithium dimethyl cuprate to **129** by Smith.[206] In each of these studies, the diagnostic interpretation favored SET to **129** based on the formation of the propellanone **130** as the only major product of the reduction reaction. Analogous behavior was anticipated, therefore, for the photolysis reaction of **129** in the presence of tertiary amines in alcoholic solvents, conditions favoring SET reactions.

Unexpectedly, enone **131**, the product of reductive cleavage of the tosylate group, was the major product (97%) from the photolysis of

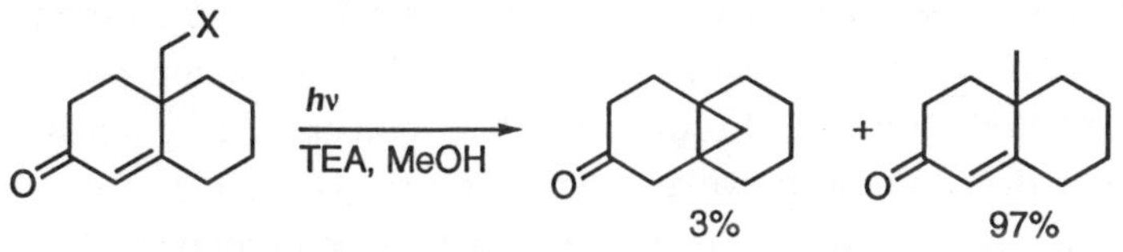

hν
TEA, MeOH
3% 97%

129 ($X = OTs, OMs, OCOCF_3, Br$) **130** **131**

Scheme 48. The SET reduction of 10-substituted octalones by triethylamine.

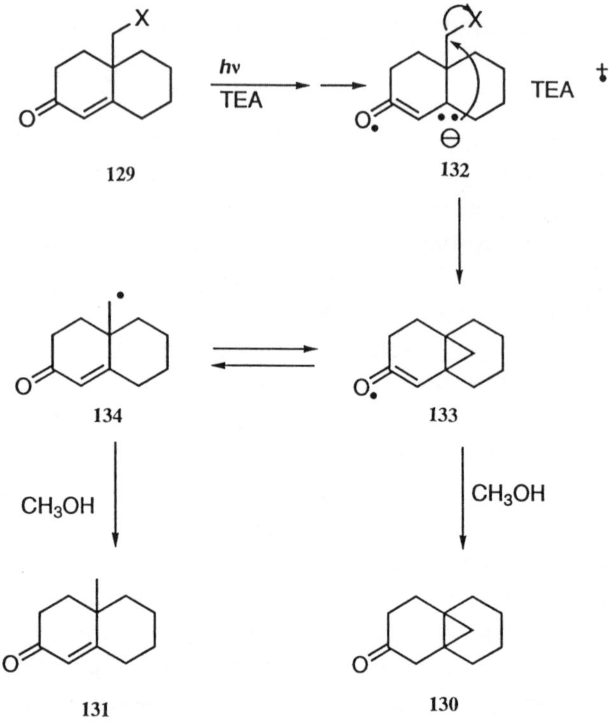

X hν TEA TEA ‡ 129 132 134 133 CH_3OH CH_3OH 131 130

Scheme 49. Givens and Atwater's[204] mechanism for the SET reduction of 10-substituted octalones by tertiary amines.

enone **129** with triethylamine (TEA) in methanol (Scheme 47).[204] Only a trace (3%) of Stork's propellanone **130** was formed. A mechanism involving the direct photoreductive cleavage of the pendant tosylate was readily dismissed by comparison of the efficiencies of a series of leaving groups having differing reduction potentials. The product distributions from three of the derivatives of **129** were identical, in accordance with the expected correspondence of the individual reduction potentials as well as the parent enone **130** reduction potential. In contrast to the product distribution, however, the quantum efficiencies for the three ketones were sensitive to the nature of the leaving group and followed a pattern of leaving-group nucleofugacities, not the leaving-group reduction potentials. These results, coupled with the known propensity of cyclopropylcarbinyl radicals to undergo ring opening, suggested alternative pathways for **130**.

The intermediate semi-enone **133** postulated for the ECE sequence of the Birch-like reductions of **129**, an α-keto cyclopropylcarbinyl radical, would be susceptible to rapid ring opening to the homoallyl radical **134**, providing an *indirect* route to the reductive cleavage of the tosylate. Furthermore, equilibration of **133** and **134** is also anticipated since the two conformationally restricted radicals are stereoelectronically arranged for such an interconversion. Neither radical appears to be favored by simple electronic factors. The rate at which an equilibrium is established, however, is neither known nor easily estimated although the unimolecular rate constants reported for the ring opening of cyclopropylcarbinyl radicals are large (ca. 10^6 s^{-1})[207] when compared with the rate constant for H-atom abstraction from methanol. The rate constant for the ring closure of **134** to **133** is expected to be smaller than the ring opening rate constant, perhaps on the order of 10^3 s^{-1}, based on the reported rate constant for ring closure of the homoallyl radical.[208] The ratio of these two estimated rate constants suggests that the equilibrium constant for **134–133** is on the order of 10^3. Product distribution, however, is more likely to be determined by the relative rates of hydrogen abstraction by **133** and **134** than by the equilibrium ratio of the two rapidly interconverting radicals.

Verification of the equilibration of **133** and **134** was independently established by the synthesis of the two bromoketone precursors **129** (X = Br) and **135** in order to examine their relative reactivities with hexaphenylditin [$(Ph_3Sn)_2$; Scheme 50]. Mixtures of benzene and diethyl ether (as the H-atom source) gave an extrapolated equilibrium ratio for **134–133** of ca. 7.0 starting with either of the bromoketones.[204] This result is strongly suggestive of a facile radical interconversion of **133** and **134** as depicted in Scheme 50. These model studies do indeed support a general ECE reaction sequence as modified by a sequential cyclopropylcarbinyl–homoallyl equilibration process for the chemical reaction step. In fact, the photochemical electron transfer model poses intriguing questions regarding the nature of the intermediates in the other one-electron transfer model reactions discussed earlier. Are the same short-lived SET intermediates generated in the metal-mediated reduction and the electrochemical reduction of **129**? At present, we have no definitive answer to this question.

Attempts to extend the Stork enone model to other Birch-like reductions of α,β-unsaturated ketones have proven to be less conclusive. With carvone **136**, an enone of historic significance in [2+2] photocycloaddition chemistry,[209] a relatively uncomplicated photoreduction process was

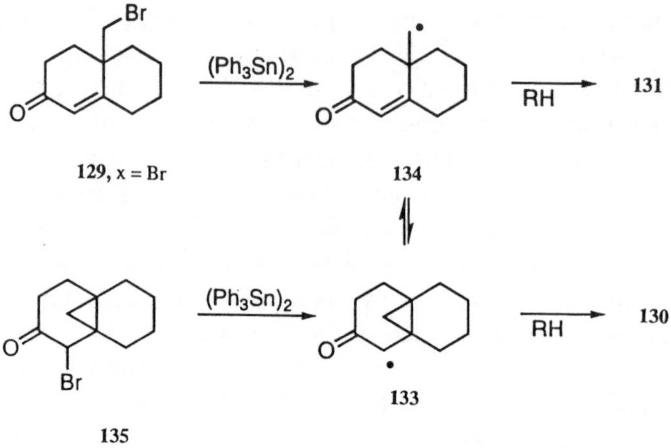

Br
$(Ph_3Sn)_2$
RH
131
129, x = Br
134
135
$(Ph_3Sn)_2$
RH
130
133

Scheme 50. Formation of radical intermediates in the SET reduction of 10-substituted octalones by tertiary amines.

demonstrated for the photolysis in TEA–methanol.[210,211] In the absence of the amine, the excited triplet of **136** yields a characteristic intramolecular [2+2] cycloaddition product (e.g., carvone camphor **138**). Addition of triethylamine to the reaction media at first results in a simple diminution of the [2+2] reaction, but this gradually gives way to the formation of a second product, dihydrocarvone **137**, as a mixture of *cis* and *trans* isomers. At a triethylamine concentration of 3.0 M, the formation of carvonecamphor **138** is completely quenched and the dihydrocarvones **137** dominates the product mixture, reaching maximum yields at 2–3 M of the amine. As the TEA concentration is further increased, even the formation of **137** is diminished, producing an overall nonlinear, bell-shaped dependence for both the disappearance of carvone **136** and the appearance of dihiydrocarvone **137** (Scheme 51).

This complex interplay between the quenching of cycloaddition and enone reduction by tertiary amines was further probed by addition of conventional triplet quenchers (e.g., piperylene or naphthalene). Each product gave an independent Stern–Volmer quenching dependency, interpreted as evidence for two separate and chemically distinct quenchable triplet precursors for the two products.[211] The Stern–Volmer slopes of 123 M^{-1} from quenching of dihydrocarvone formation and 35 M^{-1} from carvone camphor formation indicate that there are two triplets, which have lifetimes that differ by a factor of four. The shorter-lived

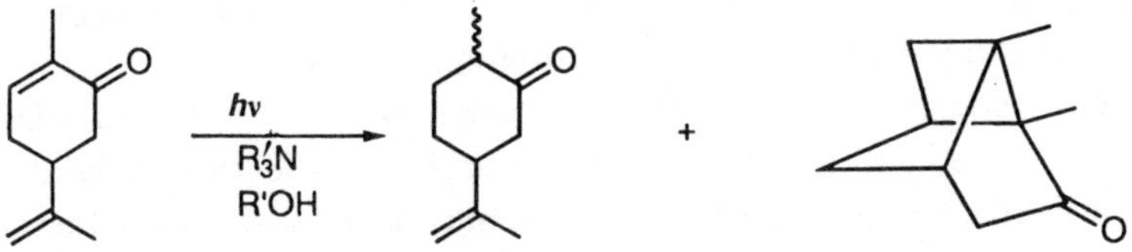

$h\nu$, R'_3N, R'OH

136 → 137 + 138

$R' = CH_3$; CD_3

R'_3N = TEA, DEABCO; PMP(CH_3); PMP (CD_3)

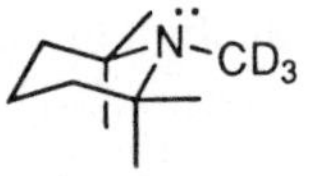

139 PMP (CD_3)

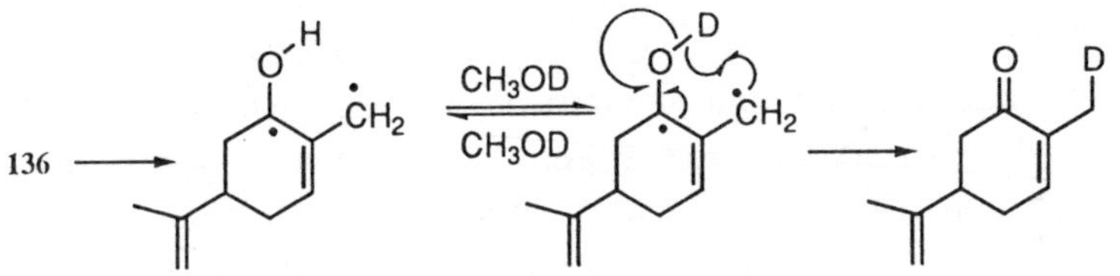

136 → 140-OH ⇌ (CH_3OD) 140-OD → 136 α-CH_2D

Scheme 51. Product distributions from the SET reduction of α,β-unsaturated enones and from the corresponding Birch reduction.

precursor to dihydrocarvone may be an enone–triethylamine exciplex, which was estimated to have a lifetime of 1.9 ns.[210] A triplet lifetime of 7.5 ns can be assigned to the enone triplet precursor to carvone camphor.

Additional evidence for a bimolecular stoichiometry for the reduction process was gleaned from a more quantitative examination of the effect of the amine concentration. A double-reciprocal plot of the efficiency for formation of **137** versus the amine concentration (at less than 2 M amine) was linear and gave a slope of 209 ME^{-1} and an intercept of 82 E^{-1}. The latter value suggested a limiting efficiency of 0.012 for the reduction process. This is substantially lower than the limiting efficiency of 0.57 derived for bicyclic enone **129** (see above). Electron backtransfer and less efficient hydrogen abstraction may account for the lower efficiency. In any event, it would appear that in the photolysis of carvone the tertiary amine is more effective as a quencher than as an electron transfer source for the photoreduction reactions.[210]

Nevertheless, the photoreactions of carvone provided valuable mechanistic information about the electron transfer process and the subsequent ground-state reactions of the resulting semi-enone. Studies employing deuterium-labeled amines and deuterated methanol have been useful for determining the proton source in the ECE reduction sequence. In a test of whether the amine served the dual function of an electron donor as well as a proton source, labelled *N*-trideuteromethyl-2,2,6,6-tetramethylpiperidine **139** ($PMPCD_3$) was employed as the electron donor in the methanol photolysis of **136**. No incorporation of deuterium was detected in any of the products nor in the recovered enone by ^2H or ^{13}C NMR or by GC-MS. Similarly, no deuterium incorporation was found for photolyses in CD_3OH where TEA, dabco, or PMP served as the electron donor.

For TEA photolyses in CH_3OD or CD_3OD, however, up to five deuterium atoms per molecule were incorporated in the isomeric *cis*- and *trans*-dihydrocarvones **137**. Figure 2 illustrates a typical deuterium distribution for *trans*-**137** as a function of the reaction conditions. A similar deuterium distribution was also found for the *cis*-**137**. Two of the deuterium atoms in carvone camphor and in the starting carvone were shown to be the result of a ground-state, base-catalyzed exchange of the α-methylene protons by the amine. A third deuterium atom was incorporated at the α-methyl group of carvone by an unusual reversible 1,5-hydrogen atom abstraction.[212] The 1,5-diradical intermediate possessing an exchangeable hydroxy proton provides a pathway for a back 1,5-deuterium transfer to yield the labelled methyl group in **137**.[211]

^1H and ^2H NMR located the fifth and final deuterium atom on **137** at the β-carbon. The introduction of the proton appears to be stereoelectronically controlled, resulting from an axial protonation of the anion center at the β-carbon of the semi-enone. This intriguing result nicely parallels the analogous stereoselectivity observed in many Birch reductions of α,β-unsaturated ketones.[201,213] In fact, the ratio of axial to equatorial attack of the proton on the carvone semi-enone was found to

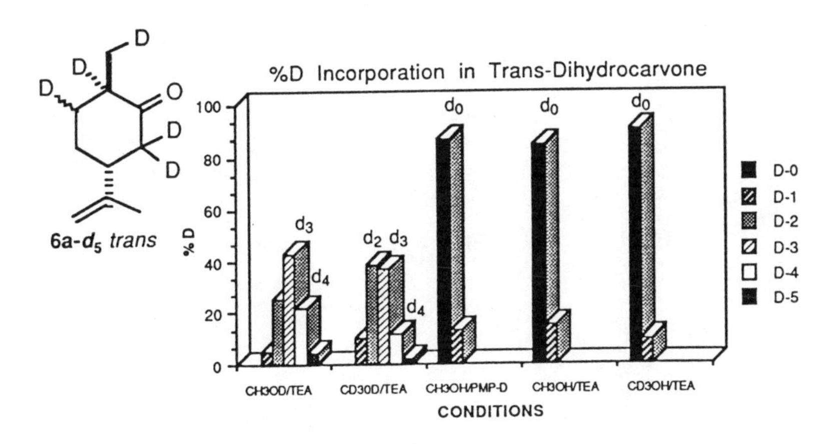

Figure 2. Percent deuterium incorporated in *trans*-dihydrocarvone.

exceed 3:1, in accordance with results from many examples of the Birch reductions of monocyclic enones.[201]

We have attempted to test the generality of these two SET reactions and to explore the overall stereoselectivity of the photoinduced electron transfer methodology with a select series of α,β-unsaturated ketones. The results have not been as encouraging, however. Photolysis of 3,4-diphenylcyclohex-2-en-1-one **141**[214] in methanol with TEA, for example, gave only a 6% yield of *cis*-3,4-diphenylcyclohexanone along with 94% of a mixture of pinacol dimers as the major product (Scheme 52). Birch reduction of the same enone also gave predominantly the *cis* isomer (34% *cis* and 2% *trans*), but at a much higher yield and with very little pinacol formation.[215] For this enone, the predominance of the *cis* isomer was attributed to steric hindrance by the 4-phenyl group of the approach of the proton donor.[213]

Likewise, photolysis of phenanthrenone **142** also gave very little reduction (9%) of the double bond.[210,211] The *cis* isomer was favored over the *trans* by a factor of 3, in stark contrast this time to a metal–ammonia reduction reported to yield exclusively the *trans* dihydrophenanthrone (*trans*-**143**) in 70% yield.[216] The major product in the photochemical reaction was, once again, a mixture of pinacols.[219] As was reported for carvone, the photorearrangement reaction, in this case a Type-A process to yield the bicyclo[3.1.0] ketone **144**, was quenched by added amine. Finally, tetracyclic enone **146** is photoreduced to the saturated analog, a process also followed by bicyclic enone **147**.

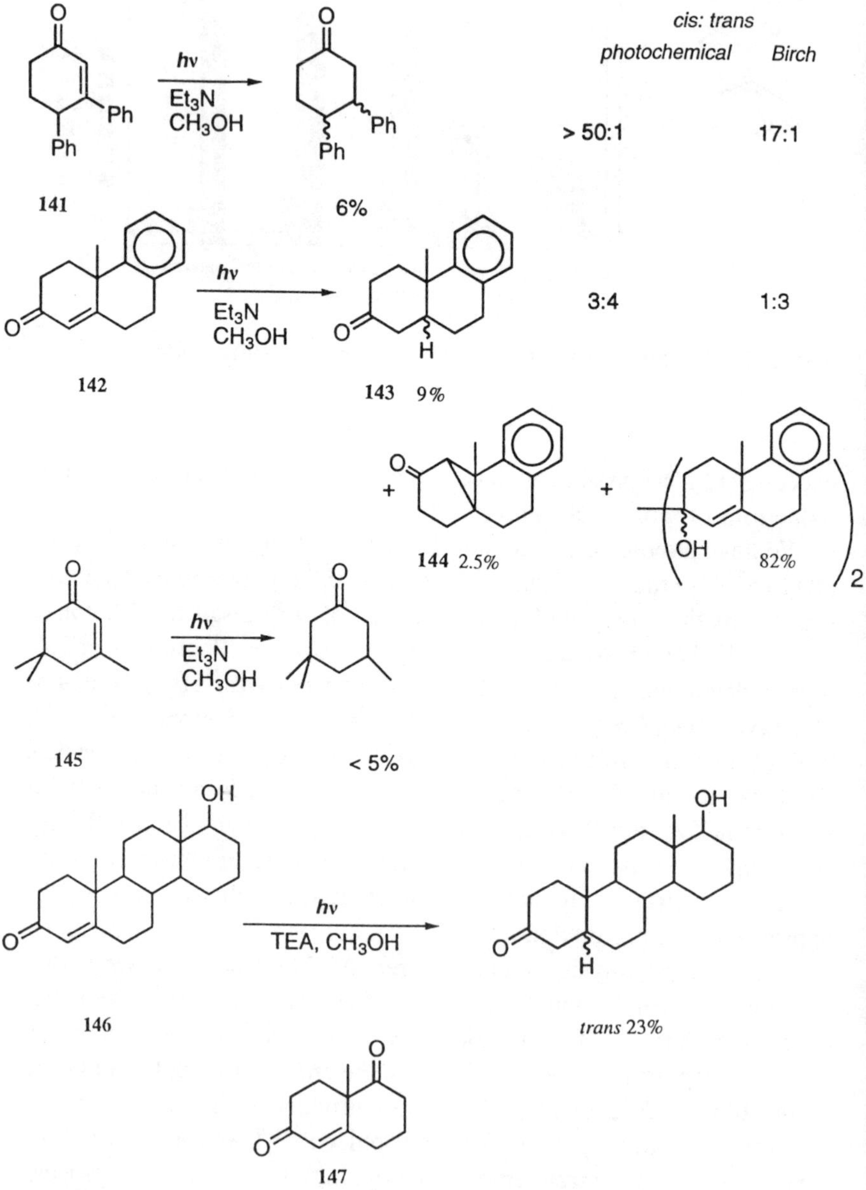

O

Ph

Ph

hν

Et_3N

CH_3OH

O

Ph

Ph

cis: trans

photochemical *Birch*

> 50:1 17:1

141 6%

hν

Et_3N

CH_3OH

O

O

H

3:4 1:3

142 **143** 9%

O

\+ +

OH

144 2.5% 82%

2

O

hν

Et_3N

CH_3OH

O

145 < 5%

OH

hν

TEA, CH_3OH

O

OH

O

H

146 *trans* 23%

O

O

147

Scheme 52. SET photoreduction of carvone with tertiary amines.

ACKNOWLEDGMENTS

Financial support for the research efforts which serve as the basis for a portion of the work described in this chapter derived from the Korea Science and Engineering Foundation (International Cooperative Research-1990 and CBM of POSTEC) to the first author, the National Science Foundation (CHE-17725 and INT-17290) and National Institutes of Health (GM-27225) to the second author, and Petroleum Research Fund administered by the American Chemical Society and the University of Kansas General Research Fund to the third author.

REFERENCES

1. Fox, M. A.; Chanon, M., Eds., *Photoinduced Electron Transfer, Parts A–D*; Elsevier: New York, 1988.
2. Fox, M. A. *Adv. Photochem.* **1986**, *13*, 295.
3. Kavarnos, G. J.; Turro, N. J. *Chem. Rev.* **1986**, *86*, 401.
4. Davidson, R. S. *Adv. Phys. Org. Chem.* **1983**, *19*, 1.
5. Lablanche-Combier, A. *Bull Soc. Chim. Fr.* **1972**, *12*, 4791.
6. Gordon, M.; Ware, W. R., Eds. *The Exciplex*; Academic Press: New York, 1975.
7. Mattes, S. L.; Farid, S. In *Organic Photochemistry*; Padwa, A., Ed.; Marcel Dekker: New York, 1983.
8. Mariano, P. S., Ed. *Advances in Electron Transfer Chemistry*; JAI: Greenwich, CT.
9. Julliard, M.; Chanon, M. *Chem. Rev.* **1983**, *83*, 425.
10. Chanon, M. *Bull Soc. Chim. Fr.* **1985**, *25*, 209.
11. Davidson, R. S. In *Molecular Association*; Foster R., Ed.; Academic Press: London, 1975.
12. Mariano, P. S.; Stavinoha, J. L. In *Synthetic Organic Photochemistry*; Horspool, W. M., Ed.; Plenum: London, 1983.
13. Marcus, R. A. *J. Phys. Chem.* **1968**, *72*, 891.
14. Scandola, F.; Balzani, V.; Schuster, G. B. *J. Am. Chem. Soc.* **1981**, *103*, 2519.
15. Rehm, D.; Weller, A. *Isr. J. Chem.* **1970**, *8*, 259.
16. Pienta, N. J. In *Photoinduced Electron Transfer*; Fox, M. A.; Chanon, M., Eds.; Elsevier: New York, 1988, Part C.
17. Ci, X.; Whitten, D. G. In *Photoinduced Electron Transfer*; Fox, M. A.; Chanon, M., Eds.; Elsevier: New York, 1988, Part C.
18. Chow, Y. L.; Danen, W. C.; Nelsen, S. F.; Rosenblatt, D. *Chem. Rev.* **1978**, *78*, 243.
19. Miller, L. L.; Nordblum, G. D.; Mayeda, E. A. *J.Org. Chem.* **1972**, *37*, 916.
20. Mann, C. K.; Barnes, K. K. *Electrochemical Reactions in Non-Aqueous Systems*; Marcel Dekker: New York, 1970.
21. Yoshida, J.; Isoe, S. *Tetrahedron Lett.***1987**, *28*, 6621.
22. Cooper, B. E.; Owen, W. J. *J. Organometal. Chem.* **1971**, *29*, 33.
23. Yoshida, J.; Maekawa, T.; Murata, T.; Matsunaga, S.; Isoe, S. *J. Am. Chem. Soc.* **1990**, *112*, 1962.
24. Yoshida, J.; Murata, T.; Isoe, S. *Tetrahedron Lett.* **1986**, *27*, 3373.

25. Bock, H.; Kaim, W. *Acc. Chem. Res.* **1982**, *15*, 9.
26. Yoshida, J., Murata, T.; Isoe, S. *J. Organometal. Chem.* **1988**, *C23–C27*, 345.
27. Lewis, F. D.; Zebrowski, B. E.; Correa, P. E. *J. Am. Chem. Soc.* **1984**, *106*, 187.
28. Murov, S. *Handbook of Photochemistry*; Marcel Dekker: New York, 1973.
29. Coyle, J. D. In *Synthetic Organic Photochemistry*; Horspool, W. H., Ed.; Plenum: New York, 1984, Chapter 4.
30. House, H. O.; Umen, M. J. *J. Am. Chem. Soc.* **1972**, *94*, 5495.
31. Schuster, D. I. In *The Chemistry of Enones*; Patai, S.; Rappoport, Z., Eds.; Wiley: New York, 1989, Chapter 15, p. 673.
32. Dunn, D. A.; Schuster, D. I.; Bonneau, R. *J. Am. Chem. Soc.* **1985**, *107*, 2802.
33. Schuster, D. I.; Dunn, D. A.; Heibel, G. E.; Broun, P. B.; Rao, J. M.; Woning, J.; Bonneau, R. *J. Am. Chem. Soc.* **1991**, *113*, 6245.
34. Fessenden, R. W.; Neta, P. *J. Phys. Chem.* **1972**, *76*, 2857.
35. Malatesta. V.; Ingold, K. U. *J. Am. Chem. Soc.* **1973**, *95*, 6400.
36. Das, S.; von Sonntag, C. *Z. Naturforsch* **1986**, *416*, 505.
37. Nelsen, S. F.; Ippoliti, J. T. *J. Am. Chem. Soc.* **1986**, *108*, 4879.
38. Dinnocenzo, J. P.; Banach, T. E. *J. Am. Chem. Soc.* **1989**, *111*, 8646.
39. Parker, V. D.; Tilset, M. *J. Am. Chem. Soc.* **1991**, *113*, 8778.
40. Nicholas, A. M. P.; Arnold, D. R. *Can. J. Chem.* **1982**, *60*, 2165.
41. Davidson, R. S.; Harrison, K.; Steiner, P. R. *J. Chem. Soc. C* **1971**, 3480; Davidson, R. S.; Orton, S. P. *J. Chem. Soc. Chem. Commun.* **1974**, 209.
42. Davidson, R. S.; Steiner, P. R. *J. Chem. Soc. Perkin Trans. 2*, **1972**, 1357.
43. Wolff, M. E. *Chem. Rev.* **1963**, *63*, 55; Kovacic, P.; Lowery, M. K.; Field, K. W. *Chem. Rev.* **1970**, *70*, 639.
44. Rosenblatt, D. H.; Hull, L. A.; DeLuca, D. C.; Davis, G. T.; Weglein, R. C.; Williams, H. K. R. *J. Am. Chem. Soc.* **1967**, *89*, 1158.
45. Audeh, C. A.; Lindsay-Smith, J. R. *J. Chem. Soc. B* **1970**, 1280.
46. Lindsay-Smith, J. R.; Mead, L. A. V. *J. Chem. Soc. Perkin Trans. 2* **1973**, 206.
47. Yoshida, K. *Electrooxidation in Organic Chemistry*; Wiley: New York, 1984.
48. Wayner, D. D. M.; McPhee, D. J.; Griller, D. *J. Am. Chem. Soc.* **1988**, *110*, 132.
49. Urry, W. H.; Juveland, O. O. *J. Am. Chem. Soc.* **1958**, *80*, 3322.
50. Giese, B. *Angew. Chem. Int. Ed. Engl.* **1983**, *22*, 753.
51. Bryce-Smith, D.; Gilbert, A. *Tetrahedron* **1977**, *33*, 2459.
52. Bellas, M.; Bryce-Smith, D.; Gilbert, A. *J. Chem. Soc. Chem. Commun.* **1967**, 862.
53. Bellas, M.; Bryce-Smith, D.; Gilbert, A. *J. Chem. Soc. Chem. Commun.* **1967**, 263.
54. Bellas, M.; Bryce-Smith, D.; Clarke, M. T.; Gilbert, A.; Klunklin, G.; Krestonosich, S.; Manning, C.; Wilson, S. *J. Chem. Soc. Perkin Trans. 1* **1977**, 2571.
55. Yang, N. C.; Libman, J. *J. Am. Chem. Soc.* **1973**, *95*, 5783.
56. Pac, C.; Sakurai, H. *Tetrahedron Lett.* **1969**, 3829.
57. Davidson, R. S. *J. Chem. Soc. Chem. Commun.* **1969**, 1450.
58. (a) Bryce-Smith, D.; Gilbert, A.; Klunklin, G. *J. Chem. Soc. Chem. Commun.* **1973**, 330; (b) Sugimoto, A.; Sumida, R.; Tamai, N.; Inoue, H.; Otsugi, Y. *Bull. Chem. Soc. Jpn.* **1981**, *54*, 3500.
59. Baltrop, J. A.; Owens, R. J. *J. Chem. Soc. Chem. Commun.* **1970**, 1462.
60. Baltrop, J. A. *Pure Appl. Chem.* **1973**, *33*, 179.

61. Gilbert, A.; Krestonosich, S. *J. Chem. Soc. Perkin Trans. 1* **1980**, 2531.
62. Bryce-Smith, D.; Gilbert, A.; Krestonosich, S. *J. Chem. Soc. Chem. Commun.* **1976**, 405.
63. Gilbert, A.; Krestonosich, S. *J. Chem. Soc. Perkin Trans. 1* **1980**, 1393.
64. Bernardi, R.; Caronna, T.; Morrocchi, S.; Ursini, M.; Vittimberga, B. M. *J. Chem. Soc. Perkin Trans. 1* **1990**, 97.
65. Bernardi, R.; Caronna, T.; Morrocchi, S.; Vittimberga, B. M. *J. Chem. Soc. Perkin Trans. 1* **1991**, 1411.
66. Pac, C.; Tosa, T.; Sakurai, H. *Bull. Chem. Soc. Jpn.* **1972**, *45*, 1169.
67. Bunce, N. J.; Gallagher, J. C. *J. Org. Chem.* **1982**, *47*, 1955.
68. Miller, L. L.; Narang, R. S. *Science 169*, 368.
69. Soloveichik, O. M.; Ivanov, V. L.; Kuzmin, M. G. *J. Org. Chem. USSR Engl.* **1976**, *12*, 860.
70. Ohashi, M.; Tsujimoto, K.; Seki, K. *J. Chem. Soc. Chem. Commun.* **1973**, 384.
71. Bunce, N. J.; Pilon, P.; Ruzo, L. O.; Sturch, D. J. *J. Org. Chem.* **1976**, *41*, 3023.
72. Bunce, N. J.; Kumar, Y.; Ravanal, L.; Safe, S. *J. Chem. Soc. Perkin Trans. 2* **1978**, 880.
73. Chittim, B.; Safe, S.; Bunce, N. J. *Can. J. Chem.* **1978**, 56, 1253.
74. Bunce, N. J. *J. Org. Chem.* **1982**, *47*, 1948.
75. Davidson, R. S.; Goodin, J. W. *Tetrahedron Lett.* **1981**, *22*, 163.
76. Gilbert, A.; Krestonosich, S.; Westover, D. L. *J. Chem. Soc. Perkin Trans. 1* **1981**, 295.
77. Mattes, S. L.; Farid, S. *J. Chem. Soc. Chem. Commun.* **1981**, 138.
78. Maroulis, A. J.; Shigemitsu, Y.; Arnold, D. R. *J. Am. Chem. Soc.* **1978**, *100*, 535; Neutenfeld, R. A.; Arnold, D. R. *J. Am. Chem. Soc.* **1973**, *95*, 4080.
79. Kellet, M. A.; Whitten, D. G.; Gould, I. R.; Bergmark, W. R. *J. Am. Chem. Soc.* **1991**, *113*, 358.
80. Ohashi, M.; Myake, K.; Tsujimoto, K. *Bull. Chem. Soc. Jpn.* **1980**, *53*, 1683.
81. Yamada, S.; Nakagawa, Y.; Watabiki, O.; Suzuki, S; Ohashi, M. *Chem. Lett.* **1986**, 361.
82. Tsujimoto, Y.; Hayashi, M.; Nishimura, Y.; Miyamoto, T.; Odaira, Y. *Chem. Lett.* **1977**, 677.
83. Ohashi, M.; Kudo, H.; Yamada, S. *J. Am. Chem. Soc.* **1979**, *101*, 2201.
84. Hasegawa, E.; Brumfield, M. A.; Mariano, P. S.; Yoon, U. C. *J. Org. Chem.* **1988**, *53*, 5435.
85. (a) Holcman, J.; Sehested, K. *J. Chem. Soc. Farad. Trans. 1* **1975**, 1211; (b) Robinson, E. A.; Schulte-Frohlinde, D. *J. Chem. Soc. Farad. Trans. 1* **1973**, 707.
86. Lewis, F. D.; Petisce, J. R. *Tetrahedron* **1986**, *42*, 6207.
87. Arnold, D. R.; Du, X. *J. Am. Chem. Soc.* **1989**, *111*, 7666.
88. Lewis, F. D.; Correa, P. E. *J. Am. Chem. Soc.* **1981**, *103*, 7347.
89. Lewis, F. D.; Correa, P. E. *J. Am. Chem Soc.* **1984**, *106*, 194.
90. Tsujimoto, Y.; Hayashi, M.; Miyamoto, T.; Odaira, Y.; Shirota, Y. *Chem. Lett.* **1979**, 613.
91. Kawanisi, M.; Matsunaga, K. *J. Chem. Soc. Chem. Commun.* **1972**, 313.
92. Cookson, R. C.; deCosta, S. M.; Hudec, J. *J. Chem. Soc. Chem. Commun.* **1969**, 753.
93. Lewis, F. D. *Acc. Chem. Res.* **1986**, *19*, 401.

94. Lewis, F. D.; Reddy, C. D.; Scheider, S.; Gahr, M. *J. Am. Chem. Soc.* **1989**, *111*, 6465.
95. Lewis, F. D.; Reddy, C. D.; Scheider, S.; Gahr, M. *J. Am. Chem. Soc.* **1991**, *113*, 3498.
96. Aoyama, H.; Arata, Y.; Omote, Y. *J. Chem. Soc. Chem. Commun.* **1985**, 1381.
97. Aoyama, H.; Sugiyama, J.; Yoshida, M.; Hatori, H.; Hosomi, A. *J. Org. Chem.* **1992**, *57*, 3037.
98. Lewis, F. D.; Reddy, G. D. *Tetrahedron Lett.* **1990**, *31*, 5293.
99. Lewis, F. D.; Reddy, G. D. *Tetrahedron Lett.* **1992**, *33*, 4249.
100. Cohen, S. G.; Parola, A. H.; Parsons, G. H. *Chem. Rev.* **1973**, *73*, 141.
101. Inbar, S.; Linschitz, H.; Cohen, S. G. *J. Am Chem. Soc.* **1981**, *103*, 1048.
102. Cohen, S. G.; Guttenplan, J. B. *Tetrahedron Lett.* **1969**, 2125.
103. Davidson, R. S.; Lambeth, P. F.; Younis, F. A.; Wilson, R. J. *J. Chem. Soc. C* **1969**, 2203.
104. Padwa, A.; Eisenhardt, W.; Gruber, N.; Pashayan, D. *J. Am. Chem. Soc.* **1969**, *91*, 1857.
105. Davis, G. A.; Carapellucci, P. A.; Szoc, K.; Gresser, J. D. *J. Am. Chem. Soc.* **1969**, *91*, 2264.
106. Cohen, S. G.; Cohen, J. I. *J. Phys. Chem.* **1968**, *72*, 3782.
107. Davidson, R. S.; Lambeth, P. F. *J. Chem. Soc. Chem. Commun.* **1969**, 1098.
108. Turro, N. J.; Engel, P. *Mol. Photochem.* **1969**, *1*, 143.
109. Monroe, R. M.; Groff, R. P. *Tetrahedron Lett.* **1973**, 3955.
110. Khan, J.; Cohen, S. G. *J. Org. Chem.* **1991**, *56*, 938.
111. Guttenplan, J. B.; Cohen, S. G. *J. Am. Chem. Soc.* **1972**, *94*, 4040.
112. Ruley, A. L.; Norman, R. O. C. *Proc. Chem. Soc.* **1964**, 225.
113. Bartholomew, R. F.; Davidson, R. S.; Lambeth. P. F.; McKeller, J. F.; Turner, P. H. *J. Chem. Soc. Perkin Trans. 2* **1972**, 577.
114. Roth, H. D. *Mol. Photochem.* **1973**, *5*, 91.
115. Atkins, P. W.; McLaughlin, K. A.; Percival, P. W. *J. Chem. Soc. Chem. Commun.* **1973**, 121.
116. Seebach, D.; Davin, H. *J. Am. Chem. Soc.* **1971**, *93*, 2795.
117. Davidson, R. S. *J. Chem. Soc. Chem. Commun.* **1966**, 575.
118. Cohen, S. G.; Stein, N. M. *J. Am. Chem. Soc.* **1971**, *93*, 6542.
119. Cohen, S. G.; Chao, H. M.; Stein, N. *J. Am. Chem. Soc.* **1969**, *91*, 521.
120. Cohen, S. G.; Stein, N. *J. Am. Chem. Soc.* **1969**, *91*, 3690.
121. Bhattacharyya, K.; Das, P. K. *J. Phys. Chem.* **1986**, *90*, 3987.
122. Classen, R. A.; Searles, S. *J. Chem. Soc. Chem. Commun.* **1966**, 289.
123. Roth, H. J.; Elvaine, M. H. *Tetrahedron Lett.* **1970**, 2445.
124. Padwa, A.; Albrecht, F.; Singh, P.; Vega, E. *J. Am. Chem. Soc.* **1971**, *93*, 2928.
125. Wagner, P. J.; Ersfeld, D. A. *J. Am. Chem. Soc.* **1976**, *98*, 4515.
126. Padwa, A. *Acc. Chem. Res.* **1971**, *4*, 48.
127. Padwa, A.; Eisenhardt, W. *J. Am. Chem. Soc.* **1971**, *93*, 1400.
128. Shafer, L. G.; Peters, K. S. *J. Am. Chem. Soc.* **1980**, *102*, 7566.
129. Manring, L. E.; Peters, K. S. *J. Am. Chem. Soc.* **1985**, *107*, 6452.
130. Simon, J. D.; Peters, K. S. *J. Am. Chem. Soc.* **1981**, *103*, 6403.
131. Hayon, E. Ibata, J.; Lichtin, N. N.; Simic, M. *J. Phys. Chem.* **1972**, *76*, 2072.
132. Lilie, J.; Henglein, A. *Ber. Bunsenges.* **1969**, *73*, 170.

133. Simon, J. D.; Peters, K. S. *J. Am. Chem. Soc.* **1982**, *104*, 614.
134. Simon, J. D.; Peters, K. S. *J. Am. Chem. Soc.* **1982**, *104*, 6542.
135. Simon, J. D.; Peters, K. S. *J. Am. Chem. Soc.* **1983**, *105*, 4875.
136. Libman, J. *J. Am. Chem. Soc.* **1975**, *97*, 4139.
137. Arnold, D. R.; Maroulis, A. J. *J. Am. Chem. Soc.* **1976**, *98*, 5931.
138. Eaton, D. F. *J. Am. Chem. Soc.* **1980**, *102*, 3280.
139. Lan, J. Y.; Heuckeroth, R. O.; Mariano, P. S. *J. Am. Chem. Soc.* **1987**, *109*, 2738.
140. Borg, R. M.; Heuckeroth, R. O.; Lan, J. Y.; Quillen, S. L.; Mariano, P. S. *J. Am. Chem. Soc.* **1987**, *109*, 2728.
141. Meth-Cohn, O. *Tetrahedron Lett.* **1970**, 1235.
142. Neadle, D. J.; Pollitt, R. J. *J. Chem. Soc. C.* **1969**, 2127.
143. Davidson, R. S.; Korkut, S.; Steiner, P. R. *J. Chem. Soc. Chem. Commun.* **1971**, 1052.
144. Davidson, R. S.; Steiner, P. R. *J. Chem. Soc. C.* **1971**, 1682.
145. Davidson, R. S.; Steiner, P. R. *J. Chem. Soc. Perkin Trans. 2* **1972**, 1357.
146. Davidson, R. S.; Harrison, K.; Steiner, P. R. *J. Chem. Soc. C* **1971**, 3480.
147. Davidson, R. S.; Orton, S. P. *J. Chem. Soc. Chem. Commun.* **1974**, 209.
148. Lee, L. Y. C.; Ci, X.; Giannotti, C.; Whitten, D. G. *J. Am. Chem. Soc.* **1986**, *108*, 175.
149. Ci, X.; Lee, L. Y. C.; Whitten, D. G. *J. Am. Chem. Soc.* **1987**, *109*, 2536.
150. Ci, X.; Whitten, D. G. *J. Am Chem. Soc.* **1987**, *109*, 7215.
151. Haugen, C. M.; Whitten, D. G. *J. Am. Chem. Soc.* **1989**, *111*, 7281.
152. Kellet, M. A.; Whitten, D. G. *J. Am. Chem. Soc.* **1989**, *111*, 2314.
153. Ci, X.; daSilva, R. S.; Nicodem, D.; Whitten, D. G. *J. Am. Chem. Soc.* **1989**, *111*, 1337.
154. Ci, X.; Kellett, M. A.; Whitten, D. G. *J. Am. Chem. Soc.* **1991**, *113*, 3893.
155. Ohga, K.; Mariano, P. S. *J. Am. Chem. Soc.* **1982**, *104*, 617.
156. Ohga, K.; Yoon, U. C.; Mariano, P. S. *J. Org. Chem.* **1984**, *49*, 213.
157. Cho, I. S.; Tu, C. L.: Mariano, P. S. *J. Am. Chem. Soc.* **1990**, *112*, 3594.
158. Dinnocenzo, J. P.; Farid, S.; Goodman, J. L.; Gould, I. R.; Todd, W. R.; Mattes, S. L. *J. Am. Chem. Soc.* **1989**, *111*, 8973.
159. Todd, W. P.; Dinnocenzo, J. P.; Farid, S.; Goodman, J. L.; Gould, I. R. *J. Am. Chem. Soc.* **1991**, *113*, 3601.
160. d'Allesandro, N.; Albini, A.; Mariano, P. S. *J. Org. Chem.* **1993**, *58*, 829.
161. Cookson, R. C.; Hudec, J.; Mirza, N. A. *J. Chem. Soc. Chem. Commun.* **1968**, 180.
162. Pienta, N.; McKimmey, J. E. *J. Am. Chem. Soc.* **1982**, *104*, 5501.
163. Pienta, N. *J. Am. Chem. Soc.* **1984**, *106*, 2704.
164. Smith, D. W.; Pienta, N. J. *Tetrahedron Lett.* **1984**, *25*, 915.
165. Dunn, D. A.; Schuster, D. I.; Bonneau, R. *J. Am. Chem. Soc.* **1985**, *107*, 2802.
166. Weir, D.; Scaiano, J. C.; Schuster, D. I. *Can. J. Chem.* **1988**, *66*, 2595.
167. Schuster, D. I.; Insogma, A. M. *J. Org. Chem.* **1991**, *56*, 1879.
168. Yoon, U. C.; Kim, J. U.; Hasegawa, E.; Mariano, P. S. *J. Am. Chem. Soc.* **1987**, *109*, 4421.
169. Hasegawa, E.; Xu, W.; Mariano, P. S., Yoon, U. C.; Kim, J. U. *J. Am. Chem. Soc.* **1988**, *110*, 8099.

170. Hasegawa, E.; Xu, W.; Mariano, P. S., Yoon, U. C.; Kim, J. U. *J. Am Chem. Soc.* **1992**, *57*, 1422.
171. Yoon, U. C.; Kim, H. J.; Mariano, P. S. *Heterocycles* **1989**, *29*, 1041.
172. Hayon, E.; Ibata, J.; Lichtin, N. N.; Simic, M. *J. Phys. Chem.* **172**, *76*, 2072.
173. Lilie, J.; Henglein, A. *Ber. Bunsenges. Phys. Chem.* **1969**, *73*, 170.
174. Xu, W.; Jeon, Y. T.; Hasegawa, E.; Yoon, U. C.; Mariano, P. S. *J. Am. Chem. Soc.* **1989**, *111*, 413.
175. Xu, W.; Zhang, X. M.; Mariano, P. S. *J. Am. Chem. Soc.* **1991**, *113*, 8863.
176. Xu, W.; Mariano, P. S. *J. Am. Chem. Soc.* **1991**, *113*, 1431.
177. Porter, N. A.; Magnin, P. R.; Wright, B. T. *J. Am. Chem. Soc.* **1986**, *110*, 2787.
178. Jeon, Y. T.; Lee, C. P.; Mariano, P. S. *J. Am. Chem. Soc.* **1991**, *113*, 8847.
179. Zhang, X. M.; Mariano, P. S. *J. Org. Chem.* **1991**, *56*, 1655.
180. Chanon, M.; Eberson, L. In *Photoinduced Electron Transfer*; Fox, M. A.; Chanon, M., Eds.; Elsevier, 1988, Part A.
181. Zhang, X.; Jung, Y. S.; Mariano, P. S.; Fox, M. A.; Martin, P. S.; Merkert, J. *Tetrahedron Lett.* **1993**, *34*, 5239.
182. Castro, P.; Overman, L. E.; Zhang, X.; Mariano, P. S. *Tetrahedron Lett.* **1993**, *34*, 5243.
183. Jung, Y. S.; Mariano, P. S. unpublished results.
184. Kim, S.Y.; Mariano, P. S. *Tetrahedron Lett.* **1994**, *35*, 999.
185. Szwarc, M. *Acc. Chem. Res.* **1969**, *2*, 87.
186. Devadoss, C.; Fessenden, R. W. *J. Phys. Chem.* **1990**, *94*, 4540.
187. Devadoss, C.; Fessenden, R. W. *J. Phys. Chem.* **1991**, *95*, 7253.
188. Miyasaka, H.; Morita, K.; Kamada, K.; Mataga, N. *Chem. Phys. Lett.* **1991**, *178*, 504.
189. Haselbach, E.; Patrice, J.; Pilloud, D.; Suppan, P.; Vauthey, E. *J. Phys. Chem.* **1991**, *95*, 7115.
190. Peters, K. S.; Lee, J. *J. Phys. Chem.* **1993**, in press.
191. Wagner, P. J.; Kemppainen, A. E. *J. Am. Chem. Soc.* **1969**, *91*, 3085.
192. Miyasaka, H.; Morita, K.; Kamada, K.; Mataga, N. *Bull. Chem. Soc. Jpn.* **1990**, *63*, 3385.
193. Miyasaka, H.; Morita, K.; Kamada, K.; Nagata, T.; Kiri, M.; Mataga, N. *Bull. Chem. Soc. Jpn. 64*, 3229.
194. Westheimer, F. H. *Chem. Rev.* **1961**, *61*, 265.
195. Bell, R. P. *Faraday Symp. Chem. Soc.* **1975**, *10*, 1.
196. Wagner, P. J.; Kemppainen, A. E.; Jellinek, T. *J. Am. Chem. Soc.* **1972**, *94*, 12.
197. Saltiel, J.; Atwater, B. W. *Adv. Photochem.* **1988**, *14*, 1.
198. Miyasaka, H.; Kiri, M.; Morita, K.; Mataga, N.; Tanimoto, Y. *Chem. Phys. Lett.* **1992**, *199*, 21.
199. Nishikawa, S.; Asahi, T.; Okada, T.; Mataga, N.; Kakitani, T. *Chem. Phys. Lett.* **1991**, *185*, 237.
200. Angel, S. A.; Peters, S. *J. Phys. Chem.* **1991**, *95*, 3606.
201. For a comprehensive review of the Birch reduction, see Caine, D. *Org. React.* **1976**, *23*, 1. The term "semi-enone" describing the radical anion reduction product of an α,β-unsaturated ketone was introduced in our 1986 communication (see footnote 1 of Ref. 204).

202. Adams, R. N. *Electrochemistry at Solid Electrodes*; Marcell Dekker: New York, 1969.
203. (a) Stork, G.; Tsuji, J. *J. Am. Chem. Soc.* **1961**, *83*, 2783; (b) Stork, G.; Rosen, P.; Goldman, N.; Coombs, R. V.; Tsuji, J. *J. Am. Chem. Soc.* **1965**, *87*, 275.
204. Givens, R. S.; Atwater, B. W. *J. Am. Chem. Soc.* **1986**, *108*, 5028.
205. Gassman. P. G.; Rasmy, O. M.; Murdock, T. O.; Saito, K. *J. Org. Chem.* **1981**, *46*, 5455.
206. Smith, R. A. J.; Hannah, D. J. *Tetrahedron* **1979**, *35*, 1183.
207. Friedrick, E. C.; Holmstead, R. L. *J. Org. Chem.* **1972**, *37*, 2550.
208. Effio, A.; Griller, D.; Ingold, K. U.; Beckwith, A. L. J.; Serelis, A. K. *J. Am. Chem. Soc.* **1980**, *102*, 1734.
209. (a) Ciamician, G.; Silber, P. *Chem. Ber.* **1908**, *41*, 1928; (b) Buchi, G.; Goldman, J. M. *J. Am. Chem. Soc.* **1955**, *79*, 4741.
210. Givens, R. S.; Singh, R.; Xue, J.-y.; Park, Y.-H. *Tetrahedron Lett.* **1990**, *31*, 6793.
211. (a) Givens, R. S.; Sinyh, R., unpublished results; (b) Singh, R., Ph.D. Thesis, University of Kansas, 1991.
212. Wagner, P. *Acc. Chem. Res.* **1989**, *22*, 83.
213. Stork, G.; Darling, S. D. *J. Am Chem. Soc.* **1964**, *86*, 1761.
214. Park, Y -H., M.S. thesis, University of Kansas, 1990.
215. Malhotra, S. D.; Moakley, D. F.; Johnson, F. *Tetrahedron Lett.* **1967**, 1089.
216. Wenkert, E.; Stevens, T. E. *J. Am. Chem. Soc.* **1956**, *78*, 2318.

INDEX

J A I P R E S S

Advances in Detailed Reaction Mechanisms

Edited by **James M. Coxon**, *University of Canterbury, New Zealand*

The questions why? and how? are synonymous with childhood. In the childhood of scientific knowledge these questions recur. This series is to be a record of this adventure and quest e.e.e.ewithin the silica walled vessel; the solar environment; or the most intimate of laboratories, the human frame. The record of achievement testifies to the resourcefulness and imagination of people, the human spirit, and curiosity promising hope in the search for understanding.

The study of detailed reaction mechanisms, of how and why molecular change occurs, forms the basis of this series intended to highlight selected approaches which have led to advances.

Volume 1, Radical, Single Electron Transfers, and Concerted Reactions
1991, 186 pp. $90.25
ISBN 1-55938-164-7

CONTENTS: Introduction to Series: An Editors Foreword, *Albert Padwa*. Preface. *James M. Coxon*. Radical Kinetics and Mechanicistic Probe Studies, *Martin Newcomb, Texas A&M University*. Free Radical Reactions: Fragmentation and Rearrangement in Aqueous Solutions, *Michael J. Davis and Brice C. Gilbert, University of York*. Carbon-Centered Radicals From Amino Acids and their Derivatives, *Christopher J. Easton, University of Adelaide*. Cycloadditions of Allenes Reactions of Unusual Mechanistic Perspicuity, *William R. Dolbier, University of Florida*

Volume 2, Mechanisms of Biological Importance
1991, 292 pp.
ISBN 1-55938-505-7 $90.25

CONTENTS: Latent Reactivity: The Study of Enzyme Mechanisms and the Design of Enzyme Inactivators, *Andres D. Abell, University of Canterbury, Christchurch, NZ.* Non-Heme Dioxygenases: A Unified Mechanistic Interpretation of their Mode of Action, *Charles W. Jefford, University of Geneva, Switzerland.* Molecular Basis for the Enhancement and Inhibition of Bleomycin-Mediated Degradation of DNA by DNA

Prospects for the Bending-Corrected Rotating, *E.F. Hayes, P. Pendergast, Ohio State University and R.B. Walker, Los Alamos National Laboratory.* Variational Treatments of Reactive Scattering: Application of Negative Imaginary Absorbing Potentials and Contracted L2 Basis Sets to Calculate S-Matrix Elements, *Isidore Last and Michael Baer, Soreq Nuclear Research Center, Israel.* Choosing Body-Fixed Axes in Arrangement Channels Approaches to Reactive Scattering, *Russell T. Pack, Los Alamos National Laboratory.* Dynamics on Reactive Potential Energy Surfaces: The Hyperspherical View, Vincenzo Aquilanti, *Simonetta Cavalli, and Gaia Grossi, Universita di Perugia.* The Effect of Vibrational Adiabaticity on 3D Properties of the Cl HC1 Reaction, *Antonio Lagana, Universita di Perugia, Antonio Aguilar, Xavier Gimenez, and Jose M. Lucas, Universitat de Barcelona.* Four Atom Reactions, *David C. Clary and Julian Echave, University of Cambridge.* Effects of Potential Energy Surface Topography and Isotope Substitution in Atom-Diatom Chemical Reactions: The Cl $+H_2$ and D $+H_2$ Systems, *Shoji Takada, Ken-ichiro Tsuda, Institute for Molecular Science, Myodaji, Akihiko Ohsaki, Miyazaki University, and Hiroki Nakamura, Institute for Molecular Science, Myodaji.* Subject Index.

Volume 2 - Part B, 1994, 236 pp. $90.25
ISBN 1-55938-706-8

CONTENTS: Preface, *Joel M. Bowman.* A Quantum Scattering Study of the C1™ HC1 C1H™H Reaction: Centrifugal Sudden Hyperspherical Differential and Integral Cross Sections, Product Distributions, and Rate Coefficients, *George C. Schatz, Northwestern University, D. Sokolovski and J.N.L. Connor, University of Manchester.* Time Dependent Wavepacket Approach to Reactive Scattering Using Arrangement Decoupling Absorbing Potentials, *Daniel Neuhauser, University of Chicago, Richard S. Judson, Sandia National Laboratory, Michael Baer, Soreq Nuclear Research Center, Israel, and Donald J. Kouri, University of Houston.* Linear Algebraic Formulation of Reactive Scattering with General Basis Functions, *Gregory J. Tawa, Steven L. Mielke, Donald G. Truhlar, University of Minnesota, and David W. Schwenke, NASA Ames Research Center.* A New Look at Symmetrized Hyperspherical Coordinates, *Aron Kuppermann, California Institute of Technology.* Reduced Dimensionality Quantum Approaches to Tetraatomic Reactive Scattering, *Joel M. Bowman and Desheng Wang, Emory University.* Subject Index.